SUSTAINABLE TOURISM?
European Experiences

SUSTAINABLE TOURISM?
European Experiences

Edited by

Gerda K. Priestley
Departament de Geografia
Universitat Autònoma de Barcelona
Spain

J. Arwel Edwards
Department of Geography
University of Wales
Swansea
UK

and

Harry Coccossis
Department of Environmental Studies
University of the Aegean
Greece

CAB INTERNATIONAL

CAB INTERNATIONAL
Wallingford
Oxon OX10 8DE
UK

Tel: +44 (0)1491 832111
Fax: +44 (0)1491 833508
E-mail: cabi@cabi.org
Telex: 847964 (COMAGG G)

A catalogue record for this book is available from the British
Library.

ISBN 0 85199 100 9

Phototypeset in Photina by Wyvern Typesetting Ltd, Bristol
Printed and bound in the UK by Biddles Ltd, Guildford

Contents

Contributors

Rhoda C. Ballinger: Department of Maritime Studies and International Transport, University of Wales, Cardiff, PO Box 907, Cardiff CF1 3YP, United Kingdom

Harry Coccossis: Department of Environmental Studies, University of the Aegean, 17 Karantoni Street, GR-81100 Mytilene, Greece

Gordon Dickinson: Department of Geography and Topographic Science, University of Glasgow, Glasgow G12 8QQ, United Kingdom

J. Arwel Edwards: Department of Geography, University of Wales, Swansea, Singleton Park, Swansea SA2 8PP, United Kingdom

Arthur Morris: Department of Geography and Topographic Science, University of Glasgow, Glasgow G12 8QQ, United Kingdom

Philip Nicholls: Division of Geography, Coventry University, Priory Street, Coventry CV1 5FB, United Kingdom

Apostolos Parpairis: Department of Environmental Studies, University of the Aegean, 17 Karantoni Street, GR-81100 Mytilene, Greece

Gerda K. Priestley: Departament de Geografia, Universitat Autònoma de Barcelona, E-08193 Bellaterra, Barcelona, Spain

Gérard Richez: Institut de Géographie et d'Aménagement, Université de Provence, 29 Av. Schuman, F-13621 Aix-en-Provence, France

Rebecca Rippin: Fundación Cavanilles, Rectorado, Universidad de Alicante, E-03080 Alicante, Spain

Stefano Soriani: Dipartimento di Scienze Ambientali, Università ca' Foscari di Venezia, Calle Larga Santa Marta 2137, I-30123 Venezia, Italy

Fernando Vera: Instituto Universitario de Geografía, Universidad de Alicante, E-03080 Alicante, Spain

Gabriele Zanetto: Dipartimento di Scienze Ambientali, Università ca' Foscari di Venezia, Calle Larga Santa Marta 2137, I-30123 Venezia, Italy

Preface

In the past decade or so, with rising general awareness on environmental issues, there has been growing concern for the effects of economic development on the environment. In parallel there has been increasing evidence that environmental degradation may affect the prospects of growth and development. This feedback relationship between development and environment has led to a search for ways to mitigate the impacts of one on the other, a strategy widely referred to as sustainable development.

Tourism is probably the best example among human activities in which the linkages between environmental quality and economic prospects are evident. Experiences from around the world demonstrate that tourism has direct and indirect effects on environmental resources while at the same time there is evidence of a strong dependence of tourism on environmental quality. In spite of a wider consensus that tourism should be part of a general strategy towards sustainability there is little agreement on, or evidence of, how to achieve this. Sustainable tourism, although widely advocated, is subject to diverse interpretations – ranging from environmentally friendly forms of tourism to sustaining the tourist activity in a place. As tourism is a broader socioeconomic phenomenon with distinct spatial patterns around the world, and, as its highly localized environmental impacts are easily identified, the geographical perspective is particularly valuable in understanding the issues involved.

The book is intended to explore sustainable tourism and demonstrate its complexity, subtlety and diversity in an effort to clarify the issues involved in operational terms, a first step towards the development of a strategy for sustainable tourism.

The contributions deal with issues of sustainable tourism as they emerge from a series of European case-studies. Europe accounts for the largest share of world tourism/recreation supply and demand. European tourism destination areas, moreover, have a long history of development and include a diversity of types of destination – from areas which are natural or rural to urbanized, emerging destinations to mature resorts, some of which are experiencing problems of decline. Obviously sustainable tourism is not a uniform strategy but should reflect the particularities of each type and case. Studying such examples is essential both from a local/regional development perspective and from the point of view of the European Union, particularly in the mid-nineties when a joint tourism policy is being considered in the context of the scheduled revision of the Maastricht Treaty.

The book is the result of 4 years of collaboration among 12 academic institutions across the European Union in exchanging students in the area of geography, tourism and environmental studies, within the framework of the Erasmus programme.

The structure of the 12-chapter book is as follows:

- *The issues:* an introduction on the issue of sustainable tourism and its various interpretations setting the framework for the chapters to follow.
- *Natural and rural areas:* four chapters, on a natural park in Scotland, a peninsula in Wales, a proposed national forest in England and a forest area in the Costa Brava, Spain, hinterland. These represent the various scales of space – from smaller to larger – and diversity of concerns – from pure conservation to recreation.
- *Built environments:* four chapters, on marina developments in the United Kingdom, the development of tourism and second homes along the Catalan, Spain, coast, the mass tourist resorts of Valencia, Spain, and the northern Adriatic, Italy, coastline, focusing on the question of sustainability in built environments from various perspectives.
- *Islands:* two chapters on Mykonos, Greece, and on Corsica, France, illustrating contrasting levels of acceptance of tourism in local communities.
- *Conclusions:* one chapter, outlining some general issues by drawing together the experiences from the case-studies.

The editors are especially grateful to Jaime (Tago) Ferrán for his generous hospitality and patience, the Facultat de Filosofia i Lletres and, in particular, the Departament de Geografia of the Universitat Autònoma de Barcelona for their collaboration, and thank Dr Olivia Fox for translations, Joan Carles Llurdes and Martí Puig for their assistance with cartography and Dr Antoni Durà for typing the final version of the manuscript.

G.K. Priestley, J.A. Edwards and H. Coccossis,
Sitges, July 1995

1 Tourism and Sustainability: *Perspectives and Implications*

HARRY COCCOSSIS

Introduction

Tourism has developed rapidly over the past 40 years. By the turn of the century tourism is expected to become the world's leading single economic activity. According to most global forecasts it is likely to continue to develop in the future as more people seek opportunities for leisure and recreation away from their place of residence. Is there in fact any limit to the future expansion of tourism?

The expansion of tourism has had a profound effect on many destination areas. Encompassing a multitude of activities from transporting to feeding people, the social, economic and environmental effects of tourism are many and varied. In some areas it has revitalized local economies whilst in others it has destroyed them; in some areas it has reinforced local identity whilst in others it has destroyed customs, traditions and social relations; in some areas it has helped protect environmentally sensitive areas whilst in others it has wrought havoc with local ecosystems and local resources. Great concern has recently been expressed about the effects of human activities on environmental resources and the long-term effect of pursuing development patterns which affect the environment and, ultimately, threaten human activities. The concept of sustainability has gained much support, in this respect, as a means of re-establishing the balance between economic, social and environmental goals. Emphasis is placed on long-term objectives.

The following questions may be raised concerning tourism, given the close relationship that exists between environmental quality and tourism development. Is tourism sustainable? What are the costs and benefits of tourism development? What contribution can tourism make to the local economy, culture and environment?

© 1996 CAB INTERNATIONAL. *Sustainable Tourism? European Experiences* (eds G.K. Priestley, J.A. Edwards and H. Coccossis)

European destination areas, rich in natural and cultural resources, have a major advantage in the world tourism market. Despite the emergence of new destinations, they still account for a large share of the world tourist trade. Coastal resorts and islands, natural parks and rural areas, cities, towns and villages alike are subject to significant pressure from tourism. Different methods of coping with the problems of tourism/recreation are being used in many European towns, counties, regions and states that are constantly in search of new ideas to better integrate tourism into their present economy and its prospects for the future. The analysis of their experience in the field would serve to focus on the issue of sustainable tourism, understand the processes of interaction between tourism and the environment in its broadest sense, explore the policy issues involved and the tools and mechanisms used when promoting a policy of sustainable development.

Tourism as a Dynamic Activity

The past few decades have witnessed the substantial growth of tourism and its expansion worldwide. From 1950 to 1994 international tourism increased almost 20-fold, reaching a figure of over 528 million tourist arrivals. During the same period international tourism expenditure increased by over 30 times (Vellas, 1992). International tourists account for only a part of the total number of tourists. Although no reliable accounts are available, it is generally estimated that domestic tourists account for two to three times the number of international tourists. A large number of short-visit excursionists to recreation areas ranging from natural parks to historic towns also exist, for which there is no reliable global estimate. Several factors have contributed to the spectacular growth of tourism, notably, rising incomes and reduced working hours, increased leisure time, improvements in transport technology, lower travel costs, etc. Holidays, a luxury in the past, have gradually become a way of life – a right for inhabitants of countries with developed economies.

European states accounted for 60% of world tourism and over half of international tourism revenues in 1994, although their share is declining as a result of broader global restructuring socioeconomic processes (WTO, 1995). In the European Union (EU) total revenues for 1991 have been estimated to exceed 82 billion (thousand million) ECU and expenses reached 78 billion ECU, leading to a surplus of 4 billion ECU for member states, not including domestic tourism, which is estimated to contribute at least as much as international tourism. Recent estimates (WTTC, 1995) suggest that tourism produces in the EU as much as 1 trillion ECU per year and accounts for one out of eight jobs and 13% of the gross domestic product.

Tourism offers opportunities for developing leisure activities, cultural awareness and international exchanges as well as providing employment and income in several European regions. One-fourth of total EU revenues from international tourism come from four peripheral countries – Greece, Ireland, Portugal and Spain. The share of these four countries in international tourism increased in the 1980s but declined between 1988 and 1991 as tourism revenues grew faster in Denmark, Germany, France, the Netherlands and the United Kingdom. The economic impact of tourism is significant not only at a macroeconomic level but also at the level of regional development. Tourism is linked directly and indirectly to several sectors of economic activity, such as agriculture, fishing, construction, energy production, manufacturing and a whole range of tertiary activities (Pearce, 1989). As a result the local/regional multiplier effect of tourism is often considerable. Furthermore, tourism often offers unique opportunities to deprived areas, such as the centres of historic towns in economic decline, small islands and outlying areas, mountain rural areas, etc., some of which are crying out for development. Prospects for the continuing growth of tourism are particularly favourable in most European countries. Domestic tourism is likely to play a significant role in this context. Tourism is expected to become the largest single economic activity by the year 2000. By all forecasts international tourism is expected to grow in the long term. European areas, rich in cultural and natural resources, are likely to continue to benefit from such growth (WTO, 1991).

The Interdependence of Tourism and the Environment

Impact of tourism on the environment

As tourism has expanded the impact of tourism on the environment has become of major concern (OECD, 1980) since tourism depends on – and consumes – local natural and cultural resources. Four broad categories of tourism impacts of tourism can be identified:

- on natural ecosystems and resources – soil, fauna, flora, landscape, air;
- on the built environment, in particular the architectural heritage;
- on local societies – culture, values and attitudes, etc.;
- on local, regional and national economies.

The multiple impacts of tourism stem from its complex structure as an activity, as tourism is closely connected with several other activities such as trade, transport, construction, etc. Furthermore, the factors which influence tourism are not only economic but also sociocultural. As a result a comprehensive assessment of the benefits and costs of tourism involves significant conceptual and methodological difficulties.

Even if one considers only the environmental dimension, comprehensive assessment encounters substantial difficulties. These stem mostly from the indirect impact involving changes brought about in the attitudes and life-style of visitors and the local population, and their effect on the use of resources. A typical case is, for example, the indirect impact of tourism resulting from associated uncontrolled urban development in an area. It should be noted that it is often not the development of the tourist sector itself which constitutes the most serious problem but the absence of a rational manner in which to integrate it in the development of the area. Within this context, the impacts of tourism may be determined in:

- the natural and man-made landscape as a result of the size and scale of tourist facilities and associated development. The landscape is often a major source of tourist attraction in a destination;
- the use of the territory, the spatial organization and function of the area as a result of the development of tourist facilities and associated infrastructure and urban development, thus creating pressures on local resources – land, water, etc.;
- the structure and function of natural ecosystems as a result of the construction and operation of tourist facilities and the associated development – directly, through loss of vital area or pollution, or indirectly, for example, through noise or the presence of people.

The anticipated pressures from future tourism development are expected to be greater on natural habitats, natural resources – land, water and energy – and infrastructures – transport and waste water treatment facilities (CEC, 1993).

Coastal zones are expected to face severe problems as coastal tourism becomes increasingly popular. Coastal tourism is highly seasonal and is becoming increasingly intensive. This gives rise to a decrease in natural sites and open space, substantial changes in coastal landscapes and conflicts concerning the use of land, water and other resources. These negative effects are further exacerbated by the indirect effects of tourism-related urban development on trade, transport facilities, vacation houses, infrastructure, residences for those working in the tourism sector, etc. Pressure on coastal zones is likely to increase in the future. It is estimated, for example, that coastal tourism-related development in the Mediterranean could double in 20 years (Batisse and Grenon, 1989). By far the largest share of such growth is expected within the territory of the EU member states, already major destination areas.

Rural areas have also been found to face increasing pressures from re-creation and tourist development (Keane, 1992; Van der Straaten, 1992). These pressures may take the form of intensive recreation facilities for water or winter sports. Their impact on the natural environment is significant, as evidenced in natural parks or the Alpine regions. New development in remote areas to cater for domestic and international tourism constitutes a threat to rural ecosystems in two ways, by increasing pressure for construction or by encouraging traditional rural activities and practices which affect the management of natural resources to be abandoned.

The built and cultural environment in villages, towns and urban areas has become a major topic of interest for an increasingly larger number of tourists (Ashworth and Turnbridge, 1990; Lockhart and Drakakis-Smith, 1995). Although the cultural heritage of larger metropolitan areas has always attracted visitors and these areas have, in many cases, managed to cope with the presence of visitors (Law, 1995), smaller urban centres and villages that are predominantly historic in character face increasing environmental and functional problems such as congestion, noise, pollution, etc. Sometimes these problems lead to a transformation of local economies to tourism-dependent economies with the attendant impact on local identity and heritage.

Islands, especially small ones, are facing increasing pressures from tourism (Lozato-Giotart, 1990). Of particular concern are the dramatic demographic and economic shifts and transformations experienced over the past 40 years or so. Islands have been acknowledged to be cases of particular concern, given the interdependence of the social, economic and environmental systems of each as well as their diversity and uniqueness (Hess, 1990). In the past, global shifts have often led to islands being abandoned, with significant effects on environmental resources – a process referred to by some as 'desertification'. The recent expansion of tourism has led to an overutilization of local resources which have already been degraded through abandon (Coccossis, 1987).

The negative impacts of tourism have attracted considerable attention and continue to be a cause for concern (Michaud, 1983). Many of these impacts derive from so-called 'mass tourism', which has predominated in the past and is likely to do so in the future. This type of tourism is, in general, the least sensitive to local resources due to the intensive type of tourist development associated with it – designed to lower the total cost of accommodation and services – and the behaviour of tourists who are attracted by these low-cost options, and who are not sensitive to environmental quality issues. Additional impacts of mass tourism on the environment result from its marked concentration in time and space. Seasonality leads to peaks of demand and the overloading of local resources and infrastructures (Pearce, 1989).

However, the impact of tourism may also be positive. As a result of providing accommodation for the influx of tourists and visitors, local residents have the benefit of access to higher-order infrastructure, facilities and services, i.e. transport, waste water treatment, etc., and often improved environmental quality. In addition, through emulation and dissemination, local values and attitudes can change and communities can become more environmentally aware and place greater value on local resources and their cultural heritage. To generalize about the impacts of tourism is difficult as these depend on a set of interconnected factors, such as the state of conservation and particular characteristics of the local environment, type of tourists, the ability of local communities to manage resources, etc. Often, conditions that are in themselves similar may lead to positive or negative impacts. Recent changes in tourist demand and supply, however, suggest that the relationship between tourism and the environment is not simply one of cause – tourism – and effect – environmental impacts. There is increasing evidence that environmental quality is a necessary condition for sustaining tourism.

Effects of environmental quality on tourism

An important characteristic of the interaction between tourism and the environment is the existence of strong feedback mechanisms: tourism often has adverse effects on the quality – and quantity – of natural and cultural resources, but it is also affected by the decline in quality and quantity of such resources. Certain areas in Europe have already suffered the consequences of a tourism-associated decline in environmental quality which has then been followed by a decline in tourism. Such problems have occurred in particular in destination areas where tourism first developed and are closely linked with the dominant postwar model of 'mass' and 'package' tourism.

As environmental awareness increases, the values and attitudes of visitors and the local population change and expectations for improved environmental quality also increase. Environmental awareness in tourist demand is growing fast and has become one of the determining factors when choosing holiday destinations. Not only the quality of tourism supply – accommodation and services – but also the overall quality of environmental assets and resources have become key determinants of holiday choices. Degradation of the environment may be the cause of loss of competitiveness of a resort in the tourism market. It can also be the result of the loss of competitiveness as this can lead to further environmental deterioration.

Similarly, there is an increasing tendency to plan on micro and macro scales, offering high-quality environments, exploiting the comparative advantages and diversity of local resources, and setting higher standards for infra- and superstructural resources. Tourist, local and sector-related bodies are all well aware that tourism depends for its success on the environmental quality of the destination. Accordingly, developers for their part are encouraged to improve their environmental performance and prepare projects which cover the costs of infrastructure, landscaping and environmental conservation. Probably more than in any other industry the economic aspects – costs and benefits – of tourism are closely linked to environmental aspects – costs and benefits (Briassoulis and Van der Straaten, 1992).

One aspect of the increased environmental concern of tourists is a shift to energetic, selective, cultural and environmentally orientated – eco, soft –

forms of tourism, which have in common an interest in the protection of the natural and social environment and quality rather than quantity of services. Tourism and environmental quality are expected to become increasingly interdependent in the future. One should not, however, overestimate the importance of special types of tourist demand, such as ecotourism, which, although their overall share of the market is expected to grow, are likely to continue to constitute minority sectors. Not all special types of tourism are beneficial to the environment. The increasing demand for recreational activities, i.e. golf-courses, water-related sports facilities, etc., places increasing demands on local resources in the form of intensive development or associated effects – such as noise. In addition, the need for specialized facilities, i.e. waterglides, aquaparks, marinas, water-skiing areas, etc., increases building and the concentration of activities in the coastal zone and other environmentally sensitive areas. An associated problem is the limited access to such special activity areas available to the general public. Large investments are required, usually of public funds, for a limited number of users – these areas' special-purpose clientele. Such practices are coming increasingly under scrutiny from the point of view of a rational use of local resources, i.e. water or land.

The general public's increasing interest in environmental issues can make the private sector, the local and national authorities and the local community more sensitive to environmental problems. In fact, these changes have already been accompanied by shifts in public policy in support of integrating environmental issues into development policies.

Policy Issues Concerning Tourism and the Environment

Whilst the impact of tourism on the natural and man-made environment was recognized early on in the process, action has been rather slow in coming. Part of the problem may have been lack of understanding of the complex relationship between tourism and the environment. Another factor has been the lack of coordination of development and environmental policies. The issue of tourism impacts on the environment has been relatively well researched since the early 1980s when tourism expanded rapidly (OECD, 1980). Although significant progress has been made, understanding of the complex processes involved leaves much to be desired. Traditional disciplinary divisions have hindered multidisciplinary approaches. The multidimensional character and significant indirect effects involved in the relationship between tourism and the environment have been incorporated relatively recently.

Environmental policy has also been fragmented, addressing specific issues such as water pollution, air quality, etc. This has partly been due to administrative fragmentation, but also to the time-lag required to incorporate integrated approaches to environmental management which deal with entire ecosystems rather than with limited aspects of environmental quality. Furthermore, environmental conservation was perceived as opposed to development policy. Environmental quality was thus viewed as a trade-off to social and economic development. It is only in relatively recent times that a broader, more integrated view of development policy has prevailed linking the concept to that of environmental protection. The emergence of the concept of 'sustainable development' (WCED, 1987) marks this shift in perspective. A global policy of sustainability in which conservation of the environment is closely related to economic efficiency and social equality has gained wider acceptance. This view emphasizes efforts to link environmental protection to tourism development policies.

In spite of its conceptual – and political – appeal, the viability of sustainability meets with significant difficulties. How can social, economic and environmental gains and losses be assessed? Is optimization of the use of resources compatible with economic prosperity? When the question of spatial scale is introduced important conceptual and methodological conflicts arise. So, for example, should sustainability be sought at a global level only? Can sustainability be achieved at a local or regional level (Nijkamp *et al.*, 1991)? How can environmental gains and losses, which might occur at different spatial scales, be assessed?

The issue of sustainability in tourism has become a priority concern in Europe, although some of the issues involved have been debated for quite some time in the context of world tourism, mainly from a social, cultural and economic point of view. The Fifth Action Programme on the Environment of the European Union with the title *Towards Sustainability* identifies tourism as one of the priority sectors (CEC, 1993). At the informal meeting of the Council of Ministers of the Environment held at Santorini in May 1994, the central theme was 'Tourism and the Environment' (YPEHODE, 1994). In the *Green Paper on the Role of the Union in the Field of Tourism* (CEC, 1995), tourism is considered to be an important area for encouraging sustainable development. Despite widespread recognition of the need to seek strategies for sustainable tourism there seems to be a very wide margin of interpretation and perspective.

- Sustainable tourism can be interpreted from a sectorial point of view according to which the basic goal is the viability of tourist activity, more in the line of *economic sustainability of tourism*. As the focus of concern is tourist activity, the emphasis of such a strategy would imply strengthening, upgrading and even differentiation of the tourist product, often relying on organizational and technological solutions and innovations. Investment in infrastructure to increase capacity and improve services, 'resort beautification' programmes, the provision of new facilities, i.e. congress halls, water parks, etc., are some of the policy tools used in this context.

- A second interpretation is largely based on ecology as a sociocultural and political view, and strongly emphasizes the need for *ecologically sustainable tourism*. This is a conservationist approach by which priority should be placed on the protection of natural resources and ecosystems. In the context of environmental management, some tourism activities, usually identified as the 'soft' types, are acceptable as complementary and non-disturbing for the natural environment.
- The issue can be approached from a slightly different angle as *sustainable tourist development*, or the need to ensure the long-term viability of the tourist activity, recognizing the need to protect certain aspects of the environment. This approach, essentially based on an economic perspective, recognizes that environmental quality is an important factor of competitiveness and as such it should be protected. Protection extends over those aspects or dimensions of environmental quality which are directly involved in the development and marketing of the tourist product, usually aesthetics, monuments, cleanliness of beaches, traffic regulation, creation of reserved areas, etc.
- Another approach is based on ecologically sustainable economic development by which *tourism is part of a strategy for sustainable development* and in which sustainability is defined on the basis of the entire human/environment system. From this perspective environmental conservation is a goal of equal importance to economic efficiency and social equity. Tourism policies are integrated in social, economic and environmental policies but do not precede them. This constitutes a more balanced and integrated approach, closer to contemporary thinking on tourism.

To illustrate the differences between these different interpretations at a conceptual level, a triangle (Fig. 1.1) can be used to represent policy. Each vertex depicts an imaginary situation in which policy is exclusively dominated by one goal, economic efficiency, social equity or environmental conservation. The first interpretation of sustainable tourism can be identified as the policy area near the vertex corresponding to economic efficiency. The second interpretation, ecologically sustainable tourism, can be identified near the vertex corresponding to environmental conservation. The third is really a combination of economic efficiency and environmental conservation policies and can be identified somewhere along the side connecting these two vertices of the triangle. The last one seems to fit better in the centre of the triangle. There is no one, ideal tourism strategy. The above approaches reflect different priorities. Each one has its own merits and could be appropriate for different cases and settings, whether mature or emerging destinations, in growth or decline, natural areas or developed resorts, etc.

Although sustainability is a relatively new concept in environmental and tourism policies, monitoring tourist activity and protecting the

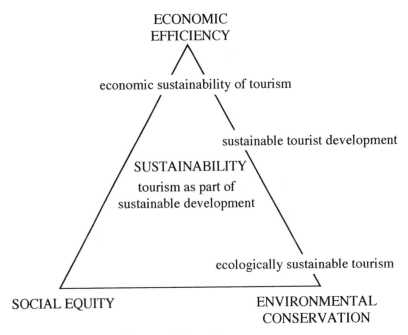

Fig. 1.1. Interpretations of sustainable tourism.

environment have been issues of policy concern for quite some time. Most policies developed to deal with the problems generated by tourism on the environment have relied on traditional environmental regulation, e.g. standards of emissions, and on physical planning tools, such as zoning and controls on urban development, which are not necessarily interrelated. In some cases, administratively established levels of tourist capacity, 'saturation' or threshold have been used, thus reflecting the tradition of planning standards. Sectorial policies on tourism, where they exist, often reflect short-term maximization of gains, underestimating the long-term effects of such strategies. Recent changes in such views incorporate elements of quality concern indirectly related to environmental issues.

Recently views on environmental policy have been enriched and expanded to include new concepts of ecosystem and resource management. This shift requires a broadening of perspectives – and tools – from issue-specific to overall regional considerations, thus affecting the way environment/tourism development issues are perceived. A wider perspective on tourism and the environment in the context of a search for strategies towards sustainable development would involve:

● linking development policy with environmental management. As a first step, the review of projects, plans and programmes from an environmental point of view should be instituted;

- regional-level environmental management schemes which would provide a framework for guiding local environmental management programmes;
- integration of tourism development and environmental management policies at the local, regional and national level;
- increasing local capacity to cope with environmental issues, particularly in rapidly developing tourist destination areas.

Priority Areas for Sustainable Tourism Development

In addition to the above actions at a general level, certain priority areas emerge as important for Europe in the immediate future (CEC, 1993; YPEHODE, 1994).

Coastal areas

Coastal areas are extremely important from an economic, social and ecological perspective. These areas of interface between terrestrial and water ecosystems include some of the most important and productive natural areas. Coastal zones also attract a wide range of economic activities.

European coastal regions constitute a unique and fragile environmental heritage. Over 50% of the ecologically richest and most sensitive areas of the EU are in coastal areas. Beaches, dunes, wetlands, estuaries, rocky shores and the coastal marine zone support a rich variety of species. Environmental quality is very important for these species and for European marine resources. Coastal areas in the EU are densely populated and heavily used. For some regions, mainly around the Mediterranean, coastal areas are of particular social and economic importance. They tend to concentrate an increasing share of population and economic activities.

Coastal areas also tend to concentrate a large number of visitors. The quality of the coastal environment is very important in providing opportunities for recreation and leisure. The combination of pleasant climate, good beaches – clear, calm waters and sandy shores – and rich cultural resources in southern regions attracts a growing number of domestic and international tourists. Tourism has contributed substantially to urban development. Infrastructure and other facilities – ports and airports, roads, waste water treatment plants, etc. – are required to cope with demand. Quite often, as the number of tourists/visitors is several times higher than the size of the local population, such investments are beyond the financial and administrative capacity of local communities. Construction and operating costs are particularly high, considering peak demand needs and the underutilization of infrastructure and facilities in the off-season period. Seasonality is an intrinsic characteristic of coastal tourism. Employment opportunities are similarly seasonal, often inducing in-migration of labour. The utilization of community services and residences is also seasonal in character, affecting social and cultural life, particularly in resorts dedicated exclusively to tourism and in very small communities.

From an economic perspective, the cost of living in tourist areas tends to be higher, as well. Property markets are also influenced by tourist demand. Of particular importance in this context is the demand for second homes or weekend residences. This leads to higher property prices and rents. It also leads to urban development and expansion of urban areas along the coasts. Estimates of the United Nations Environment Programme (UNEP) Mediterranean Action Plan – *Blue Plan* (Batisse and Grenon, 1989) – suggest that tourism and recreation could occupy 8000 km^2 of coast land by the year 2000. Coastal tourism is not unique in the Mediterranean regions. In other European coastal regions there is similar evidence of leisure-related development, although the intensity of such pressures might not be as high.

The coastal areas of Europe are facing increasing pressures for development, particularly tourist development, and are likely also to face increasing challenges to protect their environment (CEC, 1993). The size and intensity of tourist development and the associated urbanization of the coastline, the production of waste and the generation of pollution, in combination with weak planning controls, are primary causes of environmental degradation in coastal areas. A major cause, however, is the magnitude and relative scale of change, the rapid increase of tourism, which easily exceeds local capacity to cope with such change. Land-use and environmental protection policies have often been incapable of facing such challenges. Most development has occurred along the coast, which, due to geomorphological constraints, often offers the only opportunity for expansion, either as extensions of existing small settlements and towns, or in new locations at a distance.

The impacts of tourism on coastal areas include: *loss of habitat areas* due to urban development and pollution; *decline in biodiversity* of species, ecosystems and interspecies variation as a result of losses in habitat areas; *coastal erosion* as a result of a rise in sea levels, and marine engineering works to provide land or protection for tourist installations and related facilities; *loss of forest land* through urban development and forest fires, some of which are started deliberately to provide land for development; *decline in the quantity and quality of drinking-water resources*; and *sea pollution* as a result of increasing quantities of waste water.

Coastal tourism, however, is increasingly dependent on environmental quality. European Commission objectives relevant to tourism in coastal areas call for interventions on the type of tourism, the behaviour of tourists and the quality of tourist services (CEC, 1993). Significant efforts have been made to address certain immediate and medium-term needs for development control and infrastructure – waste water treatment plants, harbours for leisure craft, coastal erosion protection, etc. Not all of these fall within the scrutiny of integrated coastal management. A comprehensive means of dealing with the pressing problems of coastal tourism is required. This should take the form of promoting sustainable development and integrated management at local and regional levels, and raising public awareness. In a broader framework of coastal management, particular attention should be paid to the use of innovative tools such as environmental indicators and carrying capacity, and to the application of zoning principles and other planning controls.

Coastal tourist areas present ideal case-studies for testing sustainable development strategies given the interdependence of tourism and the environment. Several localities and regions have successfully experimented with coastal management practices in relation to tourism and sharing this experience could be important. Although a great deal of progress has been made in the development of theoretical and legal concepts or special tools and methods to promote sustainable development, very little progress has been made in the application and testing of such tools in a comprehensive and integrated strategy for the sustainable development of areas at a site, local, regional, national or supranational level.

The adoption of environmental management programmes, with the aim of reducing the pressure on local resources, is a priority in coastal tourist areas. Several aspects can be readily identified as of immediate concern: drinking-water-saving measures, the recycling of waste water, energy saving, solid waste management, etc. Attention should be given to established tourist resorts which present evidence of tourism decline associated with environmental degradation. Special 'bootstrap' operational programmes can be introduced and supported, provided they follow the principles of environmental management in the context of a strategy for sustainable development. Special care should also be taken in rapidly developing tourist areas to adopt environmentally sound tourism development policies on the basis of the capacity of environmental factors and resources to support tourist development. Efforts should focus on establishing the administrative mechanisms and capacity to cope with the problems of tourist development control as well as assistance for the development of environmental management infrastructure.

Rural areas

The prospects and problems of rural areas have been a priority issue from a social, economic and environmental point of view. In spite of geographical differences many rural areas face problems of population decline. Although, from a short-term perspective, demographic and economic decline might seem beneficial to the environment, reducing the pressure on resources at a local level, in the long run this might lead to severe environmental degradation. Often such degradation is associated with the abandon of traditional agricultural and environmental resource management practices. For example, the conversion of marginal cultivated lands to pasture, together with weak or absent resource management policies, has been responsible for the severe erosion problems faced in many mountain areas within the Mediterranean region (Coccossis, 1991). Erosion, in turn, is associated with loss of soil and vegetation, flooding of lowlands, etc.

Environmental problems in rural areas are not always attributable to abandon and neglect. Severe problems of environmental degradation are also caused by overutilization of resources in association with the absence of resource management. Tourism is by far the single most important contributor to such effects. The concentration over a relatively short period of time of large numbers of visitors in small resorts, the impacts of the construction of associated facilities and infrastructures, the waste generated, noise and traffic pollution are among the frequently cited causes of serious environmental degradation. Many natural areas and parks of Europe also suffer from this type of problems related to overcrowding. Large-scale facilities and large-scale events, noisy activities, particularly motorized recreation, i.e. water-skiing, jet boats and motor boats, four-wheel drives and motor bikes, etc., are particularly harmful to local wildlife and their habitat.

Tourism, however, also has significant beneficial effects on environmental quality in some remote areas, particularly when it contributes to environmental conservation and improvement. Some types of tourism, e.g. eco- and rural tourism, have brought the problems of rural areas to the attention of society and have contributed directly or indirectly to improved environmental management of resources. Direct effects include, for example, pressure to take action for the protection of the environment in ecologically sensitive areas, while indirect effects include raising public awareness of the need to value and protect local resources. Increased interest in soft tourism in mountain areas has also contributed to creating employment and income opportunities, and raising the standards of living in rural mountain areas.

In many respects the future of rural areas seems ambiguous within a European context, as wide-ranging socioeconomic changes would suggest the trend towards more frequent and shorter holidays more evenly distributed throughout the year, reducing the peaks of seasonal demand on local resources, but increasing the number of visitors. Structural shifts of tourist demand towards quality services encourage the adoption of environmentally based management policies. The costs and benefits of tourism are likely to spread to remote inland areas, threatening natural resources and, at the same time, offering opportunities for income, employment and better services for disadvantaged areas. Whether it will be mostly the costs or the benefits that local societies will reap from tourism will depend to a great extent on their own capabilities and resources to manage tourism properly. Tourism can contribute in a positive way to the development of rural areas but to achieve this it would certainly be necessary to put emphasis on the management of environmental resources.

Particular attention should be paid to rural ecosystem degradation directly or indirectly caused by tourism. Actions supporting tourism in rural areas should reflect environmental concern for rural ecosystems. Certain issues should be high on the priority list.

- Support should be given and emphasis placed on the preparation of environmental management programmes at local and regional levels for areas with a high risk of erosion and which face problems of abandon or intensive development due to tourism.
- Local environmental management programmes should be based on the capacity of environmental resources – land, water, habitats, etc. – to sustain human activities. Within the framework of sustainable development strategies, local environmental management programmes should adopt a long-term view of ecosystem management and should incorporate policies to manage visitor flows in time and space.
- Regional environmental management programmes should be orientated towards better integration of development plans and programmes with the management of natural ecosystems. An essential component of such environmental management programmes is education and awareness actions aimed at visitors, tourist agents and operators and the local population.
- Environmentally friendly alternative types of tourism should be introduced and supported through special action programmes at local and regional levels. Such programmes should integrate environmental protection and development policies. An essential component of such a strategy would be support for environmentally friendly traditional activities.
- Environmental assets should be identified, protected and preserved. Special priority should be given to the protection and management of natural areas of special interest. The development of leisure and recreation activities should be based on the capacity of local and regional resources to sustain such activities.
- Monitoring of local and regional environmental resources should be an integral part of any plan or programme which seeks to introduce tourism into ecologically sensitive rural areas. Local and regional administrative structures should be encouraged to reflect environmental management considerations through training to improve skills and knowledge of environmental issues.

Built environment and urban areas

The link between tourism and the environment is important for urban areas as well. Although tourism is only one of a broad spectrum of activities supported in urban areas, some villages, towns and cities face increasing problems because of tourism. Cultural resources and, in particular, the built environment have always been a major attraction for visitors. They can directly or indirectly generate tourist flows and business opportunities. Heritage, entertainment and business are the main motives for visits to urban areas. Some of these are particularly rich in such resources and obtain substantial benefits from the development of tourism. Most of the very large metropolitan areas in Europe – Paris, London, Madrid, Amsterdam, etc. – have been traditional tourist destinations. Some urban areas depend on tourism for their economic survival, especially world and European heritage cities like Venice or Bruges. Yet sometimes they suffer from the development of the activity on which they thrive. Large metropolitan areas may have sufficient administrative and financial means for coping with tourism; smaller cities may, however, encounter significant difficulties.

The increasing flow of visitors sometimes poses a threat to cultural resources such as monuments and buildings as a result of pollution or rapid degradation from frequent use. As tourism increases significant functional problems sometimes appear, e.g. congestion, noise, pollution, higher cost of living (including housing), etc., which affect the quality of life for visitors and locals alike. Often the total number of tourists in a particular time period is several times higher than the number of local people. Accommodating the flows of visitors becomes a serious problem for those urban areas which may be small in size and may have a particularly strong historic character. It is not only monuments which face threats through tourism, but also infrastructure, facilities – such as museums – and services – such as public transport, which become congested and overutilized. Waste disposal can also become a severe problem as a result of the large number of visitors in some tourist areas.

The effects of tourism on the urban environment are not always negative. Often tourism contributes substantially to the conservation of cultural resources. Public investments can be stimulated to restore and protect cultural resources. The high value placed on the built heritage by visitors can also arouse the awareness of local people to the need to protect these resources. Thus, there is a growing concern as to how to integrate tourism into the general functioning of urban historic centres. An effective local planning system is, in general, a prerequisite for accommodating tourism-related development. In recent years, however, it seems that such controls are not always adequate. Comprehensive policies are needed to integrate economic and social development with environmental protection. Often local management of tourist flows, through restricted access to vehicles, limited and controlled parking or accommodation provision, etc., can be an effective mechanism for reducing pressure from tourism. In some cases, particularly where there are large numbers of short-term or one-day visitors, such management practices are not as effective in reducing threats to and pressures on local resources. In other cases, weak administrative structures make the development of an efficient and effective management policy for tourism problems impossible.

The recent developments in tourist demand already mentioned – shorter and more frequent holidays, a qualitative shift towards alternatives to mass tourism and, in particular, soft and cultural tourism – have contributed to an expansion of urban tourism, which is expected to continue growing in the future. European historic towns will undoubtedly experience significant pressures from such growth, and their problems and prospects are not a matter of local concern alone. The problems of urban areas are the object of urban planning policy and, in this respect, the problems encountered in historic urban centres due to tourism should be integrated into local urban planning policies. These policies should be broadened to reflect the growing concern of Europeans for environmental issues. Economic development, social equity and environmental conservation should be an integral part of urban plans and programmes. However, some immediate action can be taken to relieve the pressure on and problems affecting urban historic centres as a result of tourism, until such integration of environmental management is achieved. In urban areas with a strong historical character emphasis should be placed on visitor flow management and noise control. Support for the development of environmental management programmes at local level, orientated towards reducing and controlling the flow of tourists, should be given. Local and regional plans should be encouraged to incorporate environmental management dimensions on the basis of existing infrastructure to cope with the presence of tourism.

Islands

Islands are of particular interest given the interdependence of economic, social and environmental factors. The dynamics of development/environment opportunities and constraints in the case of small islands are best exemplified in the case of tourism. The growing demand for tourism provides new opportunities, especially for small islands, providing income and employment. At the same time, however, tourism can cause damage to the island environment, culture, monuments and local social structures. Sometimes, even the quality of the visitors' experience can be affected, when the number of tourists exceeds the island's carrying capacity and functional problems arise as a result. The tremendous pressures on islands and their human and natural resources as a result of global economic restructuring processes in conjunction with the associated risks, which are extremely pronounced in the case of the small islands, demonstrate the urgency with which action should be taken in this respect.

Many islands are among the most disadvantaged areas in their respective countries. They face problems of isolation and limited market size. Often located in peripheral areas, they suffer from higher costs because of distance, lack of economies of scale, higher costs of access to information and enduring higher administrative and organizational costs and they face uncertainty, a factor which influences locational decisions. Future prospects for islands are mixed, depending on their geographical location in relation to the major centres of economic decision-making, local resources (particularly relating to tourism), the ability of island societies to mobilize local resources and the quality of their environment and services.

Environmental management acquires a central role in development policy in the context of islands, and certain issues emerge as priority considerations with regard to the management of local environmental resources, particularly in relation to tourism.

- The measurement of impacts on the stability of island ecosystems using a broad-based methodology which integrates quantitative and qualitative tools and analyses basic parameters of tourist activity, tourist presence, carrying capacity and level of integration.
- The formulation of a general assessment of the impacts of tourism on the economy and society, with particular reference to pressures on the environment, public awareness of environmental issues and sociocultural impacts on values and lifestyles.

As societies become increasingly aware of the need to reconsider the paths to development followed in the past, interest will turn to the case of islands, particularly the smaller ones, where the effects of development policies – or the lack of them – tend to be more pronounced. At an operational level, the issue of carrying capacity – or the limits to development on the basis of the island's environmental and human resources – is central to the application of sustainable development strategies on islands. Although conceptually attractive, the operational definition of carrying capacity encounters significant methodological obstacles, particularly in the case of islands where sociocultural, economic and other dimensions are associated with the environmental aspects of 'carrying capacity'.

The development of appropriate planning tools and mechanisms will be necessary if environmental, economic and sociocultural aspects are to be integrated. Monitoring tools will have to be developed which, combined with evaluation tools, will provide effective support for making policy decisions. The effectiveness of any management plan or programme depends not only on its coherence and internal logic but also on its acceptability and the commitment of all those involved. In this respect, particular emphasis should be placed on raising local, regional, national and international awareness to the particularities of islands. Priority should be placed on islands which have already been developed as major tourist destination areas. The development of local environmental management plans should be encouraged to reduce and control the level of tourist activity and improve the quality of services and the environment. Local and regional administrative structures should be strengthened to cope with environmental management.

Within the context of an integrated approach to development/environment, certain problems associated with tourism and the environment become self-evident and require priority attention. These have mostly to do with the need to meet peak demands and preserve local resources as far as possible. This involves: management of water resources through the application of new technologies and water-saving and recycling practices; management of energy resources through policies of energy conservation and development of environmentally friendly alternative technologies; management of liquid and solid waste through the application of appropriate technologies; protection of sensitive ecosystems and species through strict land development controls; visitor flow management plans and programmes to reduce the negative effects on the cultural and natural resources, encouraging the development of links between tourism and local activities, particularly those which have a marked incidence on the management of local natural resources.

Conclusions

General patterns of change suggest that some localities will have more opportunity than in previous times to participate in a wide-reaching network of exchange, through technological and institutional innovations and in response to trends towards greater economic integration. Each place must identify its role in the entire system, and this strategy will call for the development of maximum potential benefits from local assets, which, at the same time, would certainly enhance local identity. Such opportunities, however appealing from a human development perspective, also involve certain dangers, particularly for smaller localities which have fragile ecosystems. Specialization in function would not necessarily be in the long-term interest of local societies because of the possibility of disturbances assuming catastrophic proportions.

Several policy tools are available to manage the impacts of tourism and recreation. These include traditional land development controls – land use, zoning, transfer of development rights, etc.; environmental standards; traffic and tourist flow management; economic instruments; capacity limits, etc. (Stanners and Bourdeau, 1995). They can be integrated into existing – local, regional, national – or new special-purpose plans, i.e. strategic tourism plans, and programmes integrating environmental protection in economic and social development policies.

The search for sustainable tourism is certainly not a matter of the public domain alone. Other agents should also be concerned with the spirit of sharing the responsibility for sustainability. Each may have different needs, goals and priorities regarding the degree and extent to which the environment should be protected to withstand the pressure of tourism. Social, cultural, economic, organizational and institutional factors may also be involved (Coccossis and Nijkamp, 1995). Flexible forms of cooperation between the public and private domains, non-governmental organizations, local, regional and national authorities and supranational agents are required to take initiatives and develop new actions. This seems to be the challenge for modern societies: to translate sustainable tourism from theory to action.

References

Ashworth, G.J. and Turnbridge, J.E. (1990) *The Tourist-Historic City*. Belhaven, London, 296 pp.

Batisse, M. and Grenon, M. (eds) (1989) *Futures for the Mediterranean Basin: the Blue Plan*. Oxford University Press, Oxford, 279 pp.

Briassoulis, H. and Van der Straaten, J. (eds) (1992) *Tourism and the Environment: Regional, Economic and Policy Issues*. Kluwer Academic Publishers, Dordrecht.

CEC (Commission of the European Communities) (1993) *Fifth Programme on the Environment 1994–1997: Towards Sustainability*. DG Xl, CEC, Brussels.

CEC (1995) *Tourism: Green Paper on the Role of the Union in the Field of Tourism*. COM (95)97, CEC, Brussels.

Coccossis, H. (1987) Planning for islands. *Ekistics* 54 (323/324), 84–87.

Coccossis, H. (1991) Historical land-use changes in Mediterranean Europe. In: Brouwer, F., Thomas, A.J. and Chadwick, M. (eds) *Land-use Changes in Europe: Processes of Change, Environmental Transformations and Future Patterns*. Kluwer, Dordrecht, pp. 441–461.

Coccossis, H. and Nijkamp, P. (1995) *Sustainable Tourist Development*. Avebury, London, 190 pp.

Hess, A. (1990) Overview: sustainable development and environmental management of small islands. In: Beller, W., Ayala, P.G. and Hein, P. (eds) *Sustainable Development and Environmental Management of Small Islands*. Parthenon Press/UNESCO, Paris, pp. 3–13.

Keane, M. (1992) Rural tourism and rural development. In: Briassoulis, H. and Van der Straaten, J. (eds) *Tourism and the Environment: Regional, Economic and Policy Issues*. Kluwer Academic Publishers, Dordrecht, pp. 44–55.

Law, C.M. (1995) *Urban Tourism: Attracting Visitors to Large Cities*. Mansell, Poole, 202 pp.

Lockhart, D. and Drakakis-Smith, D. (1995) *Island Tourism: Problems and Perspectives*. Mansell, Poole, 240 pp.

Lozato-Giotart, J.P. (1990) *Méditerranée et tourisme*. Masson, Paris, 216 pp.

Michaud, J.L. (1983) *Le Tourisme face à l'environnement*. PUF, Paris, 234 pp.

Nijkamp, P., Van den Bergh, C.J. and Soeteman, F. (1991) Regional sustainable development and natural resource use. In: *Proceedings of the World Bank Annual Conference on Development Economics, 1990*. IBRD/World Bank, Washington DC, pp. 153–187.

OECD (Organization for Economic Cooperation and Development) (1980) *The Impacts of Tourism on the Environment.* OECD, Paris.

Pearce, D. (1989) *Tourist Development,* 2nd edn. Longman, Harlow, 326 pp.

Stanners, D. and Bourdeau, P. (eds) (1995) *Europe's Environment: the Dobris Assessment.* European Environmental Agency, Copenhagen, 616 pp.

Van der Straaten, J. (1992) Appropriate tourism in mountain areas. In: Briassoulis, H. and Van der Straaten, J. (eds) *Tourism and the Environment: Regional, Economic and Policy Issues.* Kluwer Academic Publishers, Dordrecht, pp. 86–96.

Vellas, F. (1992) *Le Tourisme.* Economica, Paris, 147 pp.

WCED (World Commission on Environment and Development) (1987) *Our Common Future.* Oxford University Press, Oxford, 267 pp.

WTO (World Tourism Organization) (1991) *Tourism to the Year 2000: Qualitative Aspects Affecting Global Growth.* WTO, Madrid, 49 pp.

WTO (1995) *Tourism in 1994 – Highlights.* WTO, Madrid, 13 pp.

WTTC (World Tourism and Travel Council) (1995) *Travel and Tourism's Economic Perspective.* WTTC, Brussels, 26 pp.

YPEHODE (Ministry of the Environment, Planning and Public Works) (1994) *Informal Council of Ministers of the Environment of the European Union.* Ministry of Environment, Planning and Public Works, Athens, 57 pp.

2 Environmental Degradation in the Countryside: *Loch Lomond, Scotland*

GORDON DICKINSON

Introduction

In common with a significant proportion of rural areas in the developed world, the Scottish Highlands, comprising the northern and western half of the country, have undergone a fundamental land-use change over the past 50 years. This change, which is based on new economic developments, has major environmental implications. Traditional primary land uses of agriculture and forestry are being replaced, in terms of importance, by use of the land for outdoor recreation and tourism. These leisure activities share use of the natural resource base of the countryside, with established primary uses such as forestry, agriculture, game management, wildlife conservation and water catchment management. Indeed throughout Highland Scotland multiple-purpose land use is usual. Furthermore, it is common for the most important rural tourism and recreation areas to share utilization of a physical resource base which, though providing excellent opportunities for development of such activities, is ecologically fragile and vulnerable to human impacts.

The Loch Lomond area in the southern Scottish Highlands provides an appropriate case-study for evaluation of impacts and their causes and the conflicts which result from such impacts. As this study shows, impacts are not only due to recreation and tourism, but also to changes in other types of land use and to variations in natural environmental conditions. Analysis of the causes of environmental impacts and degradation is important, both for the understanding of conflicts and for the resolution of these problems through proper, sustainable resource management strategies. This chapter examines the environmental problems which recent leisure activities have caused within the Loch Lomond catchment and evaluates their significance in relation to other causes of environmental change. The consequences of impacts caused by leisure activities and other land uses are considered, and it is shown that management of environmental resources for tourism and outdoor recreation must take place within an integrated framework. Only by the adoption of an integrated management system can the goal of sustainability be achieved. No less than for traditional rural land uses, such as farming and forestry, the long-term viability of outdoor recreational activities depends upon management of their resource base in a sustainable manner.

The Loch Lomond area, comprising the loch or lake and its catchment, is located to the northwest of the densely populated Midland Valley of Scotland, at its closest point no more than 30 km from Glasgow, the most populated city in Scotland. With a conurbation population of about 2 million, this is one of the largest cities in the United Kingdom (Fig. 2.1). Loch Lomond is the largest water body in Great Britain. More than 40 km long, hydrologically it comprises a complex three-basin system (Fig. 2.2). The northernmost 20 km is a narrow, fiord-like, glacial overdeepened trough, up to 200 m in depth. The middle 12 km section is shallower, typically no more than 50 m deep and studded with islands. The remaining southern part is very shallow, often no more than 10 m in depth. Here the Loch attains its maximum width of about 10 km. The limnology and associated biology of Loch Lomond are complex and, as is discussed later, the whole area has considerable conservation significance. There are conflicts in use and management of both the water body and its surrounding area. The patterns of vegetation and land use in the catchment area have been analysed by Dickinson (1994).

The general hydrological characteristics of the Loch Lomond basin have

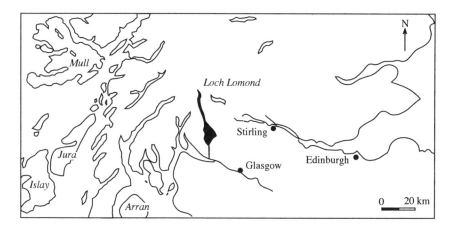

Fig. 2.1. Location of the Loch Lomond area in Scotland.

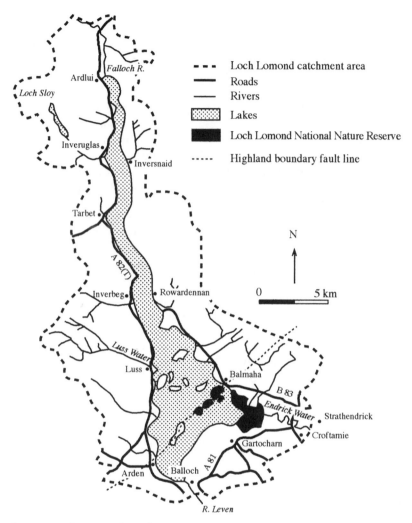

Fig. 2.2. Loch Lomond catchment area.

been described by Pender *et al.* (1993). Over half of the input of water into the Loch comes from four streams, the Falloch in the extreme north, the Luss in the west, the diverted Loch Sloy flow in the northwest and the Endrick in the south. The first two inputs are from streams with a steep gradient, draining mountains which rise to about 1000 m in height. The Sloy diversion is related to a small hydroelectric power station, and this subcatchment has a similar mountain regime. The Endrick, the largest and southernmost input,

Table 2.1. Subcatchments of the Loch Lomond catchment area (from Pender *et al.*, 1993).

Catchment	Area (km^2)	Inflow (%)
River Falloch	80.3	10.2
Luss Water	35.3	4.5
Endrick Water	219.9	27.9
Loch Sloy	80.2	10.4
Others	371.2	47.0
Total	786.9	100.0

drains the lowlands of the south and rises in the Campsie Fells, hills to the north of Glasgow, which rise to about 600 m at maximum height. The regime of this stream, though somewhat flashy, is less extreme than the first three, which have basins in which daily winter rainfall commonly exceeds 30 mm. The characteristics of the whole catchment area are shown in Table 2.1.

Tourism and Recreation in the Loch Lomond Area

The Loch Lomond area is the most popular countryside area for tourism and outdoor recreation in Scotland. A survey carried out in 1992 indicated that there were approximately 2 million visits to the area each year, the vast majority of which are made by car. The same survey indicated about 80% of visits being made by car (Dunbartonshire Enterprise, personal communication), but other studies indicate that this is probably an under-estimate, and it is likely that more than 90% of all visits are by car (Dickinson, 1988, 1995). Surveys also reveal that there is a very high concentration of tourist and recreational visits at particular times. Though there is a certain degree of leisure use of the area at all times of the year the period between mid-April and late September accounts for about 80% of visits. Furthermore, approximately 75% of trips occur on Sundays, with a further 20% on Saturdays. For locally based recreational day-trippers, visits are strongly influenced by immediate weather conditions, which can be highly variable even in summer. Thus at peak periods use of the area is in orders of magnitude greater than average figures might suggest.

Not only is there concentration of leisure use in time, but also there is clustering of activities in space. The number of locations at which visitors can park cars and gain access to the Loch shores are limited. Loch shores are by far the most popular location for sightseeing and recreational activities. Restricted access to the shores is related to the fact that most of the shore zone is in private landownership, and few private landowners permit public access to their land. Thus there is a high concentration of visits to those few sites close to the Loch which are in public ownership, and the majority of such sites are concentrated along the minor road which runs along the southern part of the eastern shore, while there are a limited number on the west side. The most popular locations have narrow shingle or coarse sand beaches and permit the launching of small boats and sailboards. Finally it should be appreciated that the spatial extent of this zone is very limited, rarely exceeding 100 m in width and commonly being much narrower. Taken together, these factors lead to a very high concentration of recreational activities at peak periods. In fragile environments such as Loch shores, site impact is related to peak levels of use rather than average levels. Although very high levels of use occur for short periods of time, it is these circumstances which have the greatest significance in generating environmental impacts and thus for resource management strategies.

The Environmental Impacts of Recreation

Over the past 25 years considerable damage to important recreational sites on the shores of Loch Lomond has been observed, notably in the form of erosion of beach and back-shore areas. During the winters of 1990, 1993 and 1994 quite severe damage was recorded around the Loch, resulting in a shoreline retreat of up to 4 m, undercutting of mature trees, necessitating their subsequent removal, and, in 1990, the collapse of a section of the east shore road. Besides the obvious environmental and ecological damage in this area of biological and landscape conservation importance, this has had significant economic consequences resulting from essential remedial and repair work.

Recent research work has focused on an evaluation of the specific role of visitor impact on the environment as a major cause of environmental degradation. This has been summarized for the Loch Lomond area by Dickinson (1993), and more generally by Selman (1992). The findings of these studies have importance not only in understanding the current situation in these environmentally vulnerable and economically important areas, but also in developing an appropriate resource management strategy for the whole catchment. Research has followed several lines. Visitor use of some of the most heavily used sites has been assessed and evidence of impacts, such as damage to vegetation and soils, investigated. Climatic and hydrological conditions during winter periods, when rapid and serious damage has resulted, have been analysed using meteorological and hydrological data. Damage resulting from these extreme conditions has been measured and surveyed immediately after the events.

In a survey of the Loch's shoreline, Tivy (1980) identified the main characteristics of shore damage. This study indicated that the most significant visitor impact on shores was located at the junction between the beach back and the vegetated terrain beyond. Damage follows the kind of sequence described by Liddle (1975). This revealed that the first species to be damaged and eliminated were herbs and small shrubs, followed by grasses, with mosses generally being the most resistant to trampling. Soil compaction in the top 5 cm resulted from foot pressure. If motor vehicles were involved, damage to vegetation was accelerated and soil impacts extended below the uppermost layer.

At Loch Lomond cars are frequently driven to the water's edge to launch boats, or simply to facilitate picnicking. As has been shown, this can only be accomplished in a few locations, which are inevitably very popular, and so here recreational impact is pronounced. This pattern is reinforced by the previously mentioned concentration of activities in time. Further recreational impacts occur through lighting fires, often at the base of mature trees, resulting in severe damage. Litter is a localized nuisance, and may be contributory to pollution problems. However, as is discussed below, at present water quality is good and such pollution problems as do occur seem to be primarily related to causes other than recreation.

However, this does not mean that there are no impacts on the water body associated with leisure activities. Two examples show that there are significant problems. A short-lived period of popularity in fishing for pike (*Esox lucius*) occurred for about 10 years from the early 1970s. This was made possible by the improved road system in Britain, which allowed anglers from northern England to reach the area for weekend trips. Pike fishing was not popular in Scotland, so stocks were good and attracted anglers from English centres. But by the early 1980s depletion of pike stocks resulted in a rapid decline in the activity. However, during this period release of the fish species ruffe (*Gymnocephalus cernuus*) into the Loch occurred. Since that time ruffe numbers have increased so rapidly that it is now the commonest fish species in the Loch, though it was not recorded as being present before the late 1970s. Ruffe is used as live bait in pike fishing, and its introduction is thus almost certainly a result of an accidental release by an angler. The drastic change in fish populations is causing concern for conservation of other fish species, notably the powan (*Coregonus cupeoides*) for which species Loch Lomond is a conservation site of international importance.

The second recreational impact results from the relatively high concentration of recreational boats in certain parts of the Loch, notably the southern basin and its islands. This spatial concentration is reinforced by the fact that most boat use is at weekends (Adams *et al.*, 1992). Launching from beaches, conflict with other recreational uses and pollution are all problems associated with boat use. Though there are serious concerns about safety as a result of the absence of controls on boat use, environmental impacts are not yet severe. However, the concentration of use may lead to the kind of problems which have been seen in the Lake District National Park in England. Because overall levels and density of use are much higher in this latter location, stringent controls on all types of boat use have been deemed necessary by resource managers.

The overall pattern of recreational impact can thus be summarized as being of limited spatial extent but of significant degree in affected sites. Research has shown that a proportion of recent environmental degradation can be accounted for by recreational activities, though other factors are involved. As is shown in the following analysis, not only are the actions of all factors crucial to the understanding of current patterns of environmental impact and change, but also all of these factors, including recreational impact, are interactive. Thus elements of environmental change and appropriate resource management strategies cannot be seen as separate entities requiring an individualistic approach; they require an integrated perspective and framework for management and conservation.

Changes in Land Use in the Loch Lomond Area

It is generally recognized that changes in land use within a catchment basin may result in environmental changes elsewhere in that catchment. In particular, basin hydrology may be modified. In the Loch Lomond area this could cause changes to the Loch itself, as well as having significant effects on the biological and landscape conservation value of the landward parts of the catchment. Over the past decade there have been considerable changes in land use. There has been a decrease in the extent of managed moorland, largely related to increases in the cost of the labour required to carry out the management of moorland by traditional 'muirburn' (Dickinson, 1994). Some of this land may return to seminatural scrub woodland, whilst other areas, particularly on the lower ground, may be given over to commercial afforestation. Coniferous plantations have greatly increased around the Loch since the 1930s. The majority of trees planted have been of the exotic species Sitka spruce (*Picea sichensis*). In the past there has been justifiable concern about the biological and landscape impact of dense plantations of single species. However, in recent years the Forestry Commission, which is responsible for most of the coniferous woodlands around Loch Lomond, has accorded outdoor recreation and conservation a high priority in their plans, so that this is now a lesser issue.

Most of the Loch Lomond catchment has Environmentally Sensitive Area (ESA) status. This central government designation qualifies areas for financial assistance towards improving land management schemes which give biological and landscape conservation a higher priority than in the past (Selman, 1992, p. 39). ESAs are located in many of the most beautiful and ecologically important areas of Britain, of which the Loch Lomond region is a good example. It is reassuring to note that the take-up of ESA schemes in the Loch Lomond area has been high, reflecting the interest of local land managers in the long-term implications of current farm and estate management practices. Examples of appropriate action which has been undertaken include support for the creation of small-scale broad-leaf plantations on farm land, and assistance with the expensive, but aesthetically important, craft of restoration of the traditional 'dry-stane dykes', unmortared stone field boundary walls constructed by hand from naturally occurring pieces of local rock.

Further changes have taken place in agricultural land use in the area. In the Strathendrick, the valley of the river draining to the southeastern corner of the Loch, there has been a continuous process of agricultural intensification for more than 100 years. This area is the most favoured agricultural part of the basin. The basis of this agriculture is sown grass, cut and conserved mainly as silage (Dickinson, 1994). Large-scale field drainage activities and high levels of nitrogenous fertilizer use are common. The former have tended to increase the peakiness of the regime of the river Endrick, whilst the latter have raised concerns about water quality. However, analyses have shown that the pattern of flow of the Endrick has little discernible effect on Loch Lomond as a whole, and monitoring has shown that generally Loch water quality is consistently very good (Dickinson *et al.*, 1991).

Water Use and Catchment Hydrology

Maintenance of water quality is important because Loch Lomond is an important source of water for domestic and industrial purposes for the populous west and central parts of the Midland Valley of Scotland, contributing about 20% of the area's total requirements. The amount of water in the Loch varies seasonally and, whilst total water available is more than adequate for present and foreseeable needs, regulation of water output helps to stabilize the level of the Loch within limits which minimize impacts on the economically important and ecologically vulnerable Loch shore zone. The detailed water quality monitoring programmes of the Central Scotland Water Development Board – the water-supply agency – and the Clyde River Purification Board – the pollution monitoring agency – confirm that even without treatment water is of very good quality (Dickinson *et al.*, 1991).

The outfall of the Loch Lomond catchment is the short south-flowing River Leven, which connects the extreme southernmost point of the Loch with the firth or estuary of the River Clyde. Outflow of water can be restricted by raising a barrage from the floor of the Leven. This is designed to come into operation when the Loch level falls below 7.92 m. This compares with a long-term average Loch level of 7.99 m for the period 1977–1990

(Dickinson *et al.*, 1991). When retracted to lie flat on the stream bed the barrage has no impact on outflow and, indeed, due to dredging to enhance flow at the barrage site, actually improves flows at times of high discharge. Thus the barrage plays a considerable role in the prevention of excessively low water levels in the Loch but has no impact on the occurrence or magnitude of high levels. The significance of this will be apparent when recent extreme climatic conditions in the Loch and their consequences for the shore zone are examined.

The Problem of Floods and Shore Erosion

Recent studies of stream flow and precipitation in the area show that there have been detectable, significant changes over the past 20 years (Dickinson *et al.*, 1991; Pender *et al.*, 1993). There has been an increase in total precipitation of at least 10% which has occurred exclusively in the winter months (October to March), with summers becoming slightly drier. In years in which winter weather has been relatively mild, runoff has further increased, since the lag imposed by winter snow has been reduced. The result has been a series of years in which the Loch has reached unprecedentedly high levels. This occurred in 1990, 1993 and 1994, in all of which years the maximum level recorded was about 1 m above the absolute maximum ever recorded over the previous 100 years. These are highly unusual circumstances. Whilst it is possible to speculate about the effects of climate change associated with global warming, and these changes are in line with those predicted for the area (Dickinson, 1991), these values are within the range which can be predicted from the actual pattern of precipitation over the past 100 years, but only just. It is unlikely that, in 3 years out of 4, such very high values should occur.

Comparison of stream flow patterns and daily Loch levels reveal that the exceptionally high levels are explained by extended periods of mild, very wet winter weather. Such weather is associated with the rapid and continuous passage of depressions over the area. These are normally accompanied by strong winds. There are limited data on wind speeds in the Loch Lomond area, but observations from Glasgow airport which is about 20 km southeast of the southern end of the Loch, confirm this. It is likely that wind speeds may be higher at most locations around the Loch, though there are microclimatic effects which modify both speeds and direction locally. The coincidence of exceptionally high Loch levels and strong winds has resulted in severe damage to the Loch shores over the past 4 years (Pender *et al.*, 1993). Calculations and observations show that waves approaching 2.5 m in amplitude can be generated in the expansive southern basin of the Loch. These waves can occur as a result of typical winter storms, in which the prevailing wind is from the southwest. The largest waves break on the southeastern shores of the Loch. As was discussed earlier, these are the most popular areas for recreation. The result of the high water levels and winter storms has been rapid and severe shore erosion in some of the most important recreational sites in the whole area. Expensive restorative work has had to be undertaken on several occasions, and in places there has been irretrievable loss of beach material and vegetated banks (Pender *et al.*, 1993).

Some protective work has been undertaken but, with the exception of those locations along which the Loch-side road runs, this is an inappropriate general response to the problem. Effective protective structures such as gabions are too expensive, and would be intrusive aesthetically. Furthermore, the erosion problem would merely be transferred to an adjacent location, probably in an accelerated form. The most appropriate response is to do as little as possible and to accept that the circumstances, whether they are a new regime or an extreme phase of the existing climatic pattern, are natural. This has very important implications for the management of the Loch's shores.

The Implications of Land Use and Environmental Changes for Resource Management

The foregoing discussion of recent human and natural environmental changes in the Loch Lomond basin has implications for the management of natural resources for tourism and outdoor recreation. It is worth reiterating that leisure activities are now the most important land use in the area, assessed by economic value, levels and extent of use or by impacts upon the environment. However, it must also be remembered that leisure activities form but one of a complementary series of land uses which make optimal use of the area's resource base. Most frequently, several uses are carried on simultaneously on a single tract of land or water. Leisure uses, to a considerable extent, depend on the aesthetic qualities of the landscape, which in no small measure are shaped by agricultural and forestry activities.

Two distinct zones for resource management can be distinguished. The shore zone, comprising beach and the vegetated beach back up to 30 m in width, is the first, and is currently subject to severe pressures. An element of this comes from recreational pressures due to high concentrations in a limited number of locations, but generally much higher impacts have resulted from high winter Loch levels, which lead to shore erosion when wind-generated storm waves attack the vulnerable shore back zone. In general, the contemporary erosion should be regarded as a natural process and accepted as part of the natural environmental regime. With the exception of a very small number of sites which have the highest human priority, e.g. where Loch-side roads cannot be satisfactorily rerouted, there is no case for engineering intervention to prevent erosion as part of a sustainable resource management strategy. Erosion should be accepted as part of the dynamic natural environmental system and, if necessary, Loch-side recreational sites such as camping areas or picnic grounds should be relocated beyond the limits of potential erosion.

The rationale for this non-interventionist approach is twofold. First, protective action is not likely to be successful. The outcome will be either a transfer of the problem to another location due to the disturbance to delicately balanced energy and sediment budgets or the breaking down of protective structures. The only possibly viable action of this type would be to 'nourish' affected beaches by adding sediment to the existing supply. Such sediment would help to absorb wave energy. However, it would have to come from a source outside the Loch's shore zone and applications would probably be necessary every year. Secondly, substantial engineering works are inappropriate to the landscape and biological conservation priorities of the area. All management work must be sympathetic to the aesthetic and scientific characteristics of the area upon which so much of its economic base depends.

The second zone is the landward part of the catchment basin, which has seen, and will continue to experience, significant land-use change. Such should be considered normal. Indeed, the present-day Highland landscapes, so highly valued, are largely a human creation of the past three or four centuries. However, as land utilization is a continuous and dynamic process, all actions, or indeed no action, will result in some change. The most important consideration at a time of rapid alterations to land-use patterns, such as the present, is what the outcome will be. Change in land use should not lead to deterioration in the sustainable resource base. Therefore conservation in the widest sense, as a part of sustainable development strategy, should be a fundamental of all land-use policy. In the recent past this has not always been the case and much land-use policy has been driven by short-term forces. Continuation of such action is likely to be environmentally damaging, economically inefficient and increasingly expensive to remedy as time passes.

At present there is very little real control over agricultural and forestry land use, though large amounts of public money go into support for these activities. There is clear evidence through the ESA schemes that voluntary actions can be successful. However, these should be backed by statutory controls to prevent damaging changes which will lead to permanent deterioration in the resource base. Current issues requiring attention in the Loch Lomond catchment include extension of intensive afforestation, construction of farm tracks and intensification of arable farming, which is based on further modification of drainage and heavy use of synthetic fertilizers. There is an understandable, if misplaced, opposition to land-use planning from landowners and managers, who fear that controls will increase bureaucracy and restrict choice. That this need not be the case has been demonstrated by the success of voluntary action. This must be extended to all parts of the area. Good planning should be less concerned with restrictions and more with working towards agreed long-term and sustainable goals.

Conclusion: the Case for Integrated Management

The solution to current problems relating to the resource base for tourism and recreation, and indeed all land uses, lies in integrated resource management. This must involve both the identification of long-term management goals and the pursuit of such goals within an overall perspective of sustainable development for the whole Loch Lomond area. The present situation is rather remote from this. Land-use policies in Scotland are generally based on a sectorial approach, which means that different sorts of land use, for example hill farming and afforestation, which have similar land resource requirements, are in competition. This may well include subventions from public funds to support the competing activities. A far better solution would be to look at the total resource base of an area and the development options for that area, and to devise an integrated scheme of management for all development undertaken. Such a scheme should not be rigid, but related to broad, agreed, long-term objectives. The general case for this approach has been convincingly argued by Mowle (1988).

There has been much interest recently in the creation of a National Park to cover the Loch Lomond area. This is not a new proposal, the idea first being conceived 60 years ago. However, concrete proposals from the Countryside Commission for Scotland (CCS), which in 1992 merged with the Nature Conservancy Council, Scotland, to form Scottish Natural Heritage, have raised the intensity of debate (CCS, 1990). Most regular visitors and many local people are broadly in favour of such a designation. This was demonstrated by evidence given to the working party set up by the government to examine the CCS proposals, and which reported in favour of the National Park proposal. However, the government remains opposed to this option, preferring a 'National Heritage Area' designation, which is little more than a continuation of the status quo. In so far as any strategy can be detected in current government policy, it seems to be based on reluctance to commit financial resources to enable the creation of a park, but depending instead on piecemeal, voluntary codes for different activities. This is short-sighted and will result both in poor resource conservation and management and in higher overall costs. The case for National Park designation has been discussed more fully by Dickinson (1991). The creation of an integrated park management structure is the simplest and most effective method of putting sustainable resource management policies in place. To safeguard the great landscape and natural resource bases of the Loch Lomond area, action is needed now. The future of tourism and outdoor recreation, which are the most important economic activities in the area, depend upon this action.

References

Adams, C.E., Tippett, R., Nunn, S. and Archibald, G. (1992) The utilisation of a large inland waterway (Loch Lomond) by recreational craft. *Scottish Geographical Magazine* 108(2), 113–118.

CCS (Countryside Commission for Scotland) (1990) *The Mountain Areas of Scotland: Conservation and Management.* Countryside Commission for Scotland, Perth, 64 pp.

Dickinson, G. (1988) Countryside recreation. In: Selman, P.H. (ed.) *Countryside Planning in Practice: the Scottish Experience.* Stirling University Press, Stirling, pp. 84–104.

Dickinson, G. (1991) National parks – Scottish needs and Spanish experience. *Scottish Geographical Magazine* 107(2), 124–129.

Dickinson, G. (1993) Environmental degradation in the Loch Lomond area: a case study of the roles of human impacts and environmental changes. In: Dawson, A.H., Jones, H.R., Small, A. and Soulsby, J.A. (eds) *Scottish Geographical Studies.* Universities of Dundee and St Andrews, Dundee and St Andrews, pp. 99–108.

Dickinson, G. (1994) Vegetation and land use in the Loch Lomond catchment. *Hydrobiologia* 290, 53–61.

Dickinson, G. (1995) Environmental impacts in the Loch Lomond area of Scotland. In: Coccossis, H. and Nijkamp, P. (eds) *Sustainable Tourism Development.* Avebury, Aldershot, pp. 159–168.

Dickinson, G., Jones, G. and Pender, G. (eds) (1991) *Loch Lomond 1991: Proceedings of the Symposium CREST.* University of Glasgow and University of Strathclyde, Glasgow, pp. 39–51.

Liddle, M. (1975) A selective review of the ecological effects of human trampling on natural ecosystems. *Biological Conservation* 7, 17–36.

Mowle, A. (1988) Integration: holy grail or sacred cow? In: Selman, P. (ed.) *Countryside Planning in Practice: the Scottish Experience.* Stirling University Press, Stirling, pp. 247–264.

Pender, G., Dickinson, G. and Herbertson, J.G. (1993) Flooding and shore damage at Loch Lomond January to March 1990. *Weather* 48(1), 8–15.

Selman, P. (1992) *Environmental Planning.* Paul Chapman Publishing, London, 195 pp.

Tivy, J. (1980) *The Effects of Recreation on Freshwater Lochs and Reservoirs in Scotland.* Countryside Commission for Scotland, Perth, 202 pp.

3 Recreation and Tourism Management in an Area of High Conservation Value: *the Gower Peninsula, South Wales*

RHODA C. BALLINGER

Introduction

Sustaining the environment of coasts of high conservation value presents a major challenge in the 1990s. Such areas, whether sandy beaches, rocky shores or spectacular cliffs along the shores of the Mediterranean Sea or the Atlantic Ocean, can attract vast numbers of tourists. The resultant pressures on the environment, often time- and place-specific, demand that sensitive management strategies are developed which balance the needs of tourism and recreation with the stringent requirements of conservation, a difficult task. This chapter provides an introduction to the geography of the recreation and tourist industry on the Gower Peninsula in South Wales (Fig. 3.1), the first Area of Outstanding Natural Beauty (AONB) to be designated in England and Wales in 1956 and also a Category V protected area as defined by the International Union for Conservation of Nature (IUCN). This designation, which also includes regional and nature parks in Spain and France, relates to seminatural landscapes in productive use. The chapter begins with an outline of the main impacts of recreational activities on the coastal environment. Particular attention is then given to the ways in which these impacts have been minimized through careful management and planning, designed to suit the stringent demands of an area of national conservation interest. Gower, being one of the earliest attempts to 'sustain' a special landscape in the UK, is a useful example to study.

In considering impacts of tourism and recreation along the Gower coast, it is necessary to view these in relation to the other pressures, which are threatening the unique ecological, cultural and landscape quality as well as the tourism potential of the Welsh and indeed the European coast. These pressures, well documented in the 1991 European Workshop on Coastal

© 1996 CAB INTERNATIONAL. *Sustainable Tourism? European Experiences*
(eds G.K. Priestley, J.A. Edwards and H. Coccossis)

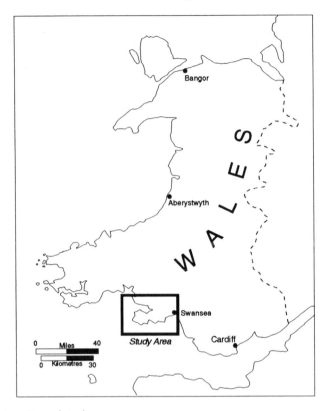

Fig. 3.1. Location of study area.

Zone Management (Countryside Commission, 1991) and in other UK reports on coastal management (NCC, 1991; RSPB, 1993), result from human, natural and organizational processes. Intense development pressures associated with urbanization and industrialization are often focused in the coastal zone. These result in increased exploitation of coastal resources, including land and offshore minerals, as well as pollution of coastal waters. Human action can also disrupt natural cycles of coastal erosion and accretion through inappropriate coastal works and from offshore dredging for aggregates. The effects of these are complicated by the uncertainty of sea-level rise and increased storminess. In addition, the complexity of the organizational and policy frameworks for addressing coastal issues is a major hindrance to the formulation of effective management solutions to many of these problems, a fact highlighted in the House of Commons Select Committee on the Environment report, *Coastal Zone Protection and Planning* (HCSCE, 1992).

As these development and other pressures become more intense, so there is a greater need for coastal management. Although the UK cannot boast a specific national coastal act like the Spanish *Ley de Costas* (Ley 22/1988), enacted because of the unrivalled pressure for onshore coastal development associated with the rapid growth of the tourist industry, there are a number of UK coastal initiatives worthy of note. The Heritage Coast (HC) programme, initiated in the 1970s by the Countryside Commission, was designed to safeguard the natural beauty and promote the recreational and amenity value of fine stretches of undeveloped coast, such as Gower. More recently the Department of the Environment and Welsh Office have issued policy guidance on planning for all coastal areas (DoE/WO, 1992a). In this, key policy issues include the conservation of the natural environment, development, particularly that requiring a coastal location and coastal risks, including flooding and erosion, as well as environmental improvements, especially of urban and despoiled coasts. The government's conservation agency in Wales, the Countryside Council for Wales, has also recently published a consultation document, *A Policy Framework for the Coastal and Marine Zone of Wales* (CCW, 1994), in which similar suggestions are voiced along with the need for a more holistic management approach to the entire Welsh coastal zone.

Case-study Description

Tourism and recreation

The complex geology of Gower has given rise to particularly varied coastal scenery, from dramatic Carboniferous limestone cliffs and sandy beaches in the south to extensive saltmarshes and dune systems along the northern shore (Fig. 3.2). Inland, the rural landscape is dominated by upland heathlands, wooded valleys and a patchwork of fields bordered by traditional stone walls. Most tourists visit Gower for this rich, unspoilt scenery (Swansea City Council, 1991) and so it is of extreme importance that this visual resource is not eroded, particularly because the tourist industry has become such a vital part of the economy of Swansea and Gower as traditional industries have declined. Tourism has become the survival route for many farmers of the peninsula, who, like many others in rural Wales, supplement their income with revenue gained from a variety of sources, including farmhouse bed and breakfast and charges for using land for camping and car parking during the summer season.

Most of the tourism infrastructure on Gower is low-key: serviced accommodation occurs in small-scale establishments, such as farmhouses and guest-houses (Fig. 3.3). In fact, the vast majority of visitors, in common with many visitors to South Wales, stay on camp or caravan sites, many of which

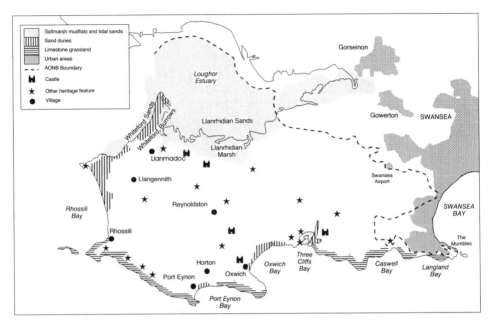

Fig. 3.2. The resource: natural and heritage features.

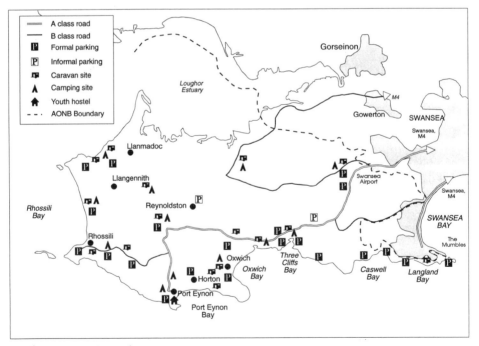

Fig. 3.3. Tourism infrastructure.

command fine sea views and are within easy walking distance of a sandy beach. The capacity of Gower to receive visitors has almost reached saturation point, however. It has been estimated that 18 million people live within 4 hours' travelling distance of the peninsula (Wilson, 1990) and recent surveys have shown that, on bank holidays and sunny summer weekends, the 10,000 or so resident population can be supplemented by up to 30,000 holiday-makers and 50,000 day-trippers from the industrial and urban areas of South Wales and western England (Bridges *et al.*, 1986). Popular beaches, such as Oxwich, receive in the region of 250,000 to 300,000 visitors each per year. It is only the severe congestion on the restricted road network of the peninsula (Fig. 3.3) that acts as a 'safety-valve' (Wilson, 1990), controlling the numbers reaching the coast at peak times. The main holiday season on Gower, however, like most of northern Europe, is confined to the school summer holidays, a 6-week period extending from the middle of July to the end of August. Midweek and for the rest of the year, one can still enjoy the relative peace and tranquillity that one might expect from an AONB.

The variety of recreational activities that take place on Gower can hardly be rivalled elsewhere in Britain (Fig. 3.4). Within a relatively short distance, the varied landscape and coastline provide opportunities for a

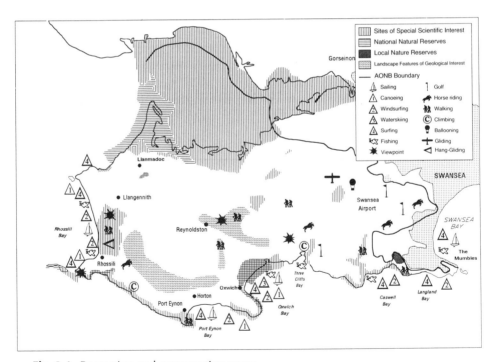

Fig. 3.4. Recreation and conservation areas.

range of water sports and land-based outdoor pursuits, from windsurfing and subaqua to rock climbing and pony-trekking. The provision of an increasing variety of attractions, including wet-weather sites, under-cover leisure activities and tourism attractions, within half an hour's drive of the peninsula, adds to Gower's intrinsic appeal. With careful marketing of this package it is hoped that the season will be extended and that a wider variety of visitors will be attracted to the area, including international tourists, particularly from the countries of northwest continental Europe (D. Nutt, Marketing and Communications Division, Swansea City Council, July 1994, personal communication). The peninsula has the potential to cater for national competitions for many activities as well as for informal car-based recreation, at present the pastime of most visitors to Gower.

Conservation

To understand the pattern and intensity of the environmental impacts of tourism and recreation on Gower, it is necessary to appreciate the management context within which these activities occur. Most of the peninsula is covered by a range of designations, which attempt to conserve its special landscapes and geological features as well as its flora and fauna (Fig. 3.4). Most of the peninsula lies within the Gower AONB, the primary purpose of which – the conservation, protection and enhancement of natural beauty – is largely accomplished through strict planning controls which restrict unsightly development. Only the limited size of the area prevented it from becoming a national park, like the Pembrokeshire Coast National Park to the west, which contains some very similar coastal scenery. Gower, like just under a third of the coastline of England and Wales, can also boast 55 km of HC, as well as a range of other designations covering nature conservation interests, including proposed possible Special Areas of Conservation (SACs) under the European Union (EU) Species and Habitats Directive (EC, 1992), 19 Sites of Special Scientific Interest (SSSIs), three National Nature Reserves (NNRs) and two Local Nature Reserves (LNRs). It is interesting to note that the objectives of the HC designation, unlike that of the AONB, include the promotion of recreational and educational opportunities so long as these are consistent with the conservation of the natural beauty and the protection of the heritage features. Table 3.1 provides a summary of the main objectives of all of these designations along with a brief guide to the legal protection which they afford.

In common with many coastlines in England and Wales, the National Trust and local County Wildlife Trust – Glamorgan Wildlife Trust – have played a major role in the environmental management of this coast. The former organization, in particular, owns and manages extensive tracts – 1295 ha – of the Gower coastline (Fig. 3.5). Much of this land was purchased under the Enterprise Neptune programme, launched in the mid-

Table 3.1. Major national conservation designations.

Designation	Objectives	Legal protection
Area of Outstanding Natural Beauty (AONB)	(i) To conserve natural beauty (ii) To promote recreation when consistent with (i) (iii) To safeguard agriculture, forestry and other rural industry and the economic and social needs of local communities	Under the National Parks and Access to the Countryside Act (1949) The Countryside Commission (CC England) and the Countryside Council for Wales (CCW), which designate these areas, advise on planning and developments affecting natural beauty
Heritage Coast (HC)	(i) Conserving the quality of coastal scenery in its natural state (ii) Encouraging enjoyment by the public, consistent with the above (i)	None Where Heritage Coasts have been indicated on the Local Authority Structure and Local Plan, they are considered in the control of development
National Nature Reserve (NNR)	Providing special opportunities for the study of, and research into, matters relating to the fauna and flora of Great Britain and the physical conditions in which they live, and for the study of geological and physiographical features of special interest *or* of preserving flora, fauna or geological or physiographical features of special interest	Under National Parks and Access to the Countryside Act (1949) NNRs are owned or leased by the English Nature (EN) or the Countryside Council for Wales (CCW)
Site of Special Scientific Interest (SSSI)	Conservation of areas not managed as a nature reserve, but which are of special interest, by reason of their flora, fauna or physiographical features	Several acts, including the Wildlife and Countryside Act (1981) Public bodies must consult with EN/CCW over applications to alter an SSSI
Local Nature Reserve (LNR)	Primarily, for the enjoyment of the general public. Secondly, for conservation of small areas – often close to urban areas – of special floral, faunal, geological or physiographical interest	Under the National Parks and Access to the Countryside Act (1949)

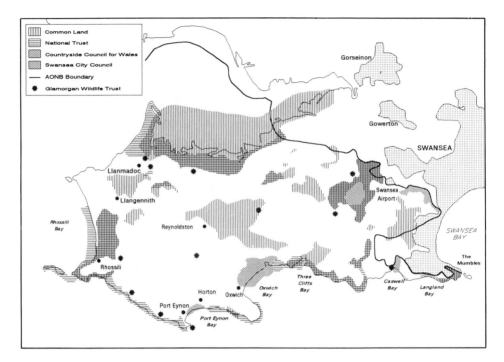

Fig. 3.5. Landownership.

1960s, which bought up stretches of unspoilt coastline in order to conserve
them for future generations. Some would argue that these sections of the
coastline are the safest from environmental degradation, as the National
Trust can declare its property 'inalienable' under a special Act of Parliament
of 1907. This means that, although inalienable land may be leased, it would
only be subject to compulsory purchase with the expressed consent of
Parliament.

Impacts of Tourism and Recreation

The impacts of recreation and tourism on the Gower environment are as
diverse as the range of habitats and recreational activities in the area. To a
large extent, the pattern of impacts is determined by the geography of the
peninsula, particularly the limited road network which controls access to
the coast. The most intense visitor usage is concentrated at a few 'honey-pot'
sites along the south coast, where extensive car parking lies adjacent to
sandy beaches (Fig. 3.3), such as at Langland Bay, Caswell Bay, Oxwich
and Port Eynon. The remoteness of areas north and west of these sites,
partly a function of the limited signposting on Gower roads, and the general

perception that the extensive saltmarshes are a less aesthetically attractive environment generally regulate visitor numbers. The following two subsections focus on the ecological impacts of recreation within the most vulnerable environments. The first deals with the impacts on the sand-dune systems of the south coast, whereas the second is concerned with disturbance to ecosystems along the rocky shores of the peninsula. The subsequent subsections focus on the pollution problems and maintenance of landscape quality in the area, essential to the longevity of the tourist and recreation industry on Gower.

Ecological impacts on sand-dunes

Through good management practice most of the potential ecological impacts of tourism and recreation have been minimized on Gower so that sites and species of high conservation value are now at little risk. Attention will therefore focus on visitor impacts on the dune systems, including the sand-dunes of the south and west coast which are of national importance because of their rare and diverse flora and invertebrate fauna. Fortunately, the most extensive dune systems of the Whiteford Burrows NNR are relatively inaccessible (Fig. 3.3). However, many of the other dune systems lie adjacent to sandy beaches, which are sought by vast numbers of tourists. A third of Europe's coastal dunes have been destroyed or seriously disfigured over the last 20 years and so sand-dunes have become a highly prized habitat within Europe and have been subject to intense management schemes. On Gower many of the erosional impacts of recreational activities at honey-pot sites have been much reduced through such careful management and foresight. Even off-road motor-bike scrambling, four-wheeled drive vehicles and other crazes which cause severe erosion in vulnerable areas have been addressed. Currently, a code of practice and route suggestions for off-road cyclists are being produced by the Gower Countryside Service (1994). The sand-dune systems at Oxwich and Port Eynon, worthy of extensive and often expensive land management, have been selected to illustrate the range of management techniques applied on Gower.

The dune system between Port Eynon and Horton has been intensely managed, because these dunes, as well as being an important habitat, protect the village of Horton from severe storms and sand encroachment. At this location, Swansea City Council, the local planning authority, has fenced off and planted marram grass on the most vulnerable sections and placed gabion mattresses alongside a stream running through the dunes. They have also installed board walks from the car parks to the beach so that visitor flows are controlled through the dune system. Interpretation boards in strategic positions inform the visitor of the purpose of the management. This scheme, so far, appears to have been successful in stabilizing the dunes. Of great concern, however, particularly for the tourist industry, is the unaccountable and dramatic fall in beach levels in front of the dune system at this site, which has occurred over the last few years. The recent reissue of a licence to abstract offshore marine aggregates has, consequently, become a major local issue.

A completely different approach is now taken at Oxwich, one of the foremost NNRs in Wales. Immediately after the Second World War, habitat management concentrated on combating the massive dune erosion which had resulted from army manoeuvres and practice D-Day landings. In the 1970s, visitors were seen as the main threat: patrolling the exclosures within the frontal dunes became a major task during the summer months. The influential study by DART (1973) during this period notes that the actual impacts of trampling had undergone little investigation, but that decline of the strand-line beetle *Eurynebria companata* might be partly attributed to this cause. The general public's awareness of the purpose of such exclusions rose over the next decade, so, by the early 1980s, a 90% stabilization of the dunes had been achieved. Unfortunately, this was accompanied by concern over a reduction in diversity of species: some floristic and entomological interest had been lost through the overfixation of sand (Hughes, undated). Today, the management of the frontal dunes attempts to maintain a balance between the three principal site objectives, namely, the maintenance of stability, the promotion of diversity and access for educational purposes. This was referred to as the 'trinamic equilibrium' by the former reserve warden. Consequently, visitors can actually make a positive contribution to management of the reserve, even if unwittingly. In sections of the rear dunes, which few visitors frequent, ponies and feral goats have been introduced to maintain the dune system.

Disturbance of ecosystems

Disturbance to elements of ecological and geological interest on the rocky shores from recreational activity is a particularly important issue along the south coast of the peninsula. Whereas the impacts and management of sand-dune systems are generally well established on Gower, the effects of rock climbing and collecting of foreshore species are more recent concerns, which have less well-defined management solutions. As such, they are worthy of note, even though these problems are confined to relatively small areas.

The spectacular, sheer Carboniferous limestone cliffs of the south coast are popular sites for rock climbing. Within the South Gower NNR south of Rhossili there are several recognized climbs, including those at Mewslade and Yellow Wall. As these lie close to nesting sites for choughs, kittiwakes and peregrines, a voluntary climbing ban has been placed on these climbs during the nesting season, from March to September. This ban, published in the *Gower and South East Wales Climbing Guidebook* and in other British mountaineering literature, generally keeps this potential impact under control (D. Paynter, Whiteford Burrows NNR, Gower, July 1994, personal communication). However, there is more concern from local CCW and National Trust wardens over the relatively new – and illegal – sport of 'bolt climbing', which could deface cliff sections of geological interest. Bolt climbing enables climbers to reach cliff sections previously inaccessible using conventional climbing techniques and equipment; at present, cliffs at Pwll Du, Three Cliffs and Lewes Castle are most at risk (D. Paynter, 1994, personal communication). Because the practitioners concentrate on inaccessible stretches of cliff, any patrol and enforcement system which might be introduced would be very difficult to implement. As the limestone cliffs of south Gower have been recently proposed as possible SACs by the UK nature conservation agencies under the EU Species and Habitats Directive contributing to the EU ecological network, Natura 2000, the maintenance of the conservation interest along this stretch of coast is of paramount importance.

Recreational angling is a popular activity throughout the year on Gower, with fishing marks such as those on Worm's Head and the Rhossili 'ledges' being favourites for the locals as well as the visitors (Fig. 3.4). The effects of the continuous removal of edible, shore and velvet crabs for bait from the rocky foreshore have only recently become a recognized issue. Frequently, undersized edible crabs are taken and the collectors do not distinguish between species, so threatening some of the rarer types (Davies, 1989). However, it appears that the apparently innocuous practice of overturning pebbles looking for crabs may have a profound effect on the microhabitat of the underside of pebbles, desiccating species once overturned (D. Paynter, 1994, personal communication). The ecology of the lower littoral zone is at considerable risk, particularly in easily accessible stretches of rocky shore, such as the causeway between the headland at Rhossili and the Worm's Head. The effects of casual rock-pool-hunting tourists and the collecting fishermen, however, are small compared with those caused by the vast and ever increasing number of educational field parties. In 1994, a Gower Outdoor Network was established, which hopes to control this problem. This will produce a code of conduct for the ecological study of rocky shores and will suggest alternative sites for school use. It is hoped that the adoption of this code of conduct by local schools and field-study centres will provide an indirect and informal means of policing the activities of the recreational users.

Pollution problems

Clean beaches and bathing water are essential to the UK coastal tourist industry. As tourists begin to expect a high quality environment, many rural coastal areas are actively promoting the cleanliness of their coastal waters and are in fear of negative publicity from the media when beaches fail the EU Bathing Water Directive (EC, 1976). The issue of bathing-water quality is very much in the public eye in South Wales, with frequent reports on the subject in the local press. Periodically, the waters in question are those along the southeastern coast of the Gower and the notoriously polluted Swansea Bay. Tourism is not entirely to blame; the main culprit is a tidally controlled, short-sea outfall off The Mumbles, from which the Lower Swansea Valley and the city of Swansea dispose of sewage effluent, which has only received preliminary treatment in the form of screening (NRA, 1994). Under certain tidal and weather conditions, this discharge and its associated combined sewer overflows cause regular non-compliance with the mandatory bacterio-logical standards of the EU Bathing Water Directive (EC, 1976) for some of these bathing waters.

Water-based recreational users occasionally come into contact with sewage-related debris along southeast Gower and reports of 'sewage sick-ness' occasionally hit the headlines. Despite the largely inconclusive results of the first major epidemiological survey in the UK at Langland Bay (Kay and McDonald, 1986) into the health risks incurred from swimming in sea water, the local branch of the action group Surfers against Sewage have continued to be most vociferous in their disapproval of the practice. How-ever, they continue to use this site despite ear, nose and throat infections and other minor disorders, apparently related to the poor water quality (Nelson, 1993). Hopefully, by 1997, Welsh Water plc, the water company responsible for sewage treatment and disposal, will have completed its pro-posed sewage treatment scheme and will be discharging effluent which has undergone tertiary treatment and ultraviolet disinfection through a long-sea outfall into Swansea Bay (NRA, 1994). From the hydrographic models pro-duced for this scheme, it is expected that the water quality, particularly in Swansea Bay, will show a marked improvement. As a result, the Director of Tourism Marketing and Communications for Swansea City Council envis-ages that, by the end of the decade, the extensive sandy beach of Swansea Bay, which is hardly used at present, will provide a realistic and more access-ible alternative to many of the Gower beaches (D. Nutt, 1994, personal communication).

West of The Mumbles, there are a few much smaller outfalls, catering for the sewage from the small rural population of south Gower. These out-falls normally give rise to little concern as the bathing waters of the south and western Gower meet the Guideline Standards laid down in the EU Bath-ing Water Quality Directive (EC, 1976). Some beaches along this coast have even gained European Blue Flag – Caswell Bay – and Keep Wales Tidy Cam-paign Seaside Awards – Port Eynon – in recent years. These take into account the Guideline and Mandatory Bacteriological Standards respectively. However, the treatment facilities at Port Eynon and Rhossili are to be upgraded by 2005 in line with the other major EU directive, which is driving the clean-up of coastal waters even further, namely, the EU Urban Waste-water Directive (EC, 1991).

Many visitors to Gower might be forgiven for thinking that litter is not a major problem. This is largely thanks to the extensive daily beach-cleaning operations, which Swansea City Council and some of the local landowners organize and fund throughout the summer season on some of the main tourist beaches. Marine, as opposed to tourist-derived, debris can be a problem along southwestern beaches, particularly where limited access and remoteness combine to preclude mechanical beach cleaning (Nelson, 1993), such as at Rhossili, where one might expect a pristine environment. Above the beach, on the Rhossili 'ledges', litter is a major issue for local conservationists (D. Paynter, 1994, personal communication). Some of the anglers who frequent this area discard litter and redundant fishing tackle, including hooks and monofilament line (Davies, 1989), creating a hazardous environment for birds and other wildlife. The approach of the Countryside Council for Wales (CCW) in dealing with this is to recruit some of the anglers as CCW Voluntary Wardens. As such they enjoy various rights and use of equipment, but, at the same time, are charged with patrolling and clearing litter (D. Paynter, 1994, personal communication).

Visual intrusion

In an AONB, the maintenance and even enhancement of the visual environment is of paramount importance. Strict development control along such coasts has prevented extensive and inappropriate linear development, the blight of tourist developments along many other sections of the European coast. In the Gower, serviced accommodation is generally low-key and small-scale. With the move towards 'green tourism' on Gower, and in accordance with central government and Wales Tourist Board guidelines on 'appropriate development' (DoE/WO, 1992a, b; WTB, 1994), existing agricultural buildings are to be converted to provide further accommodation in an unobtrusive way. This may even add to the scenic quality of some areas where derelict or poorly maintained buildings have spoilt the view.

The dual role of the Gower HC in conserving scenic beauty and in facilitating access for recreational activities, however, is often difficult to accomplish in practice. Even the most passive recreational activity can change the appearance of an area: the presence of picnickers on a beach can add colour and detract from the naturalness. Generally, the quality of scenery on Gower is as good as, if not better than, it was when it was declared an AONB over 30 years ago. However, there has been some erosion of the scenery in the main honey-pot sites. In common with many other stretches of the Welsh coast (WTB, 1994), the extensive caravan and chalet sites, poorly landscaped car parks and inadequate tourist facilities have been some of the main threats to the landscape. These are a particular intrusion where there is no screening and little natural cover in the areas of wide, open topography: this is the case of the caravan and chalet sites from Hillend to Llanmadoc, which are easily visible from the Worm's Head several kilometres away, regarded as being one of the most, if not the most, spectacular viewpoints on Gower. Fortunately, Swansea City Council have made an Order under Article 4 of the Town and Country Planning Development Order bringing these areas, which would not otherwise require planning permission, under the control of the local authority, as their permission is now required for use as temporary (touring) caravan and camping sites for 28 days per year.

However, the Director of Marketing and Communications for Swansea City Council expresses concern about the poor environmental quality of a number of major Gower destinations (D. Nutt, 1994, personal communication). The appearance of some car parks and their associated facilities is a major issue in several locations, notably at Oxwich and Rhossili, where privately owned car parks lie adjacent to areas of special landscape interest and to land managed by the CCW and the National Trust respectively. At Oxwich, the foreshore is dominated by a number of shacks which provide refreshment and toilet facilities. In contrast, sites now owned and managed by the local authority, Swansea City Council, such as those at Port Eynon and Caswell Bay, show that the impact of amenity provision can be greatly reduced through careful landscaping and rationalization.

However, in any area of special landscape quality, even measures to conserve the natural environment can come under scrutiny in terms of their impact on the visual environment. The contrasting approaches of the various conservation organizations result in different styles, although these have all been carefully designed using natural materials, such as wood and stone. Interpretation panels are also being very sensitively designed and provide the visitor with a wealth of useful information on the local environment. Unfortunately, many of these are a target for vandals, particularly in popular night-time barbecue areas along the southeast coast, such as Port Eynon, so overnight they can become prominent eyesores. The construction of paths, necessary to ensure safety and reduce footpath erosion, can also intrude into the naturalness of the landscape. At Rhossili, the commendable work of the National Trust in providing safe access to the Worm's Head has been criticized by some conservationists. A significant percentage of the public do not use the steps down to the Causeway, steps which were necessary because of severe erosion and drainage problems. Consequently, trampling here has produced a small scar because the 'desire lines', the preferred routes of the walkers, do not coincide with the new pathway. Fortunately, the wide open grassland on the top of the headland is very hardy and, having such a high carrying capacity, can cope with the vast numbers of visitors.

Managing for Sustainable Tourism?

The strength of the land-use planning system and the determined efforts of the various conservation bodies at site level have combined to ensure that the Gower is still one of the most beautiful and valued coastal landscapes in Britain. The development plans for the area, the West Glamorgan Structure Plan (Morgan, 1993) and the Swansea Local Plan (Wilson, 1995) aim to achieve a balance between the pressures for tourism, recreation and the other activities on Gower in accordance with the principles of positive conservation (Mullard, 1995). Within the structure plan, three categories of administrative zones – remote, intermediate and intensive – have been defined. In remote sites, such as Rhossili, there has been a presumption against development of any kind, whereas in intermediate zones – Oxwich, Pennard – although the presumption still generally holds, limited recreational and service facilities are allowed. In intensive zones – Port Eynon, Caswell – appropriate visitor facilities are permitted and have actually been encouraged to improve the coastal landscape. These designations have been seen to be a major guide to development control and a major influence on the development of tourism and recreational facilities on the Gower since the mid-1980s. These, along with site-level zoning policies on some NNRs, such as at Oxwich, where entry is restricted in the most sensitive and valuable conservation areas, have ensured that west Gower is relatively well protected from mass tourism. Other areas, however, have become targets for tourism and recreational development, although these tend to be in the more urbanized southeast, which has long since been modified from its natural state.

The local authority administrative zones are incorporated into the Gower Management Plan (Wilson, 1990), which provides a 'framework for the conservation and general improvement of the area, coordinating individual schemes into a concerted strategy for action'. This is being put into practice and developed by the Gower Countryside Service, led by a Gower countryside officer with responsibility for the Gower AONB and HC, the first senior appointment to an AONB in the UK (Mullard, 1995). One of the keys to the success of the plan and service has been the coordination which has been achieved in a relatively short time; the Gower Countryside Forum, including representatives from the conservation, tourism and recreational sectors, has been established and holds regular meetings, including site visits, throughout the year. The exchange of information is vital to understanding the complex relationships between the sectors and is essential for the formulation of effective sustainable development strategies.

A new tourism strategy, incorporating many of the principles of environmentally sustainable tourism listed in the Wales tourism document *Tourism 2000* (WTB, 1994), is currently being developed by the Gower Countryside

Service and Swansea Tourism Forum (D. Nutt, 1994, personal communication). This will promote activity and special-interest holidays compatible with sustainable tourism, developing attractions based on the natural characteristics of Gower. In the short-term, out-of-season bird-watching holidays, cycle routes and a series of circular walks are being developed. The promotion of public transport to the Gower, particularly for some of these new activity-based breaks will also be encouraged. One small step towards this has been the use of a local bus service for the Gower special events over the summer months. Attempts to provide a shuttle bus service, like those operating in the Lake District and Snowdonia National Parks, however, has been thwarted by fears from local bus companies that the service would not be economic. In addition to the schemes outlined above, more effective interpretation and more selective marketing of the conservation area will be attempted. The use of small-scale visitor information panels is currently being piloted around the peninsula along with the development of low-key visitor information points, often housed in village shops.

The success of these new measures, however, may be overshadowed by a number of external factors, including the completion of the second Severn crossing in 1996 and the projected increase in car ownership from 20 to 25 million by 2000 (WTB, 1994). There are also pressures from other sectors, including dormitory housing developments for Swansea in eastern Gower, and the problem of the unexplained drop in beach levels and loss of sand over recent years.

With over 4000 full-time job equivalents in the tourist industry and a total revenue from tourism spending in the Swansea, Mumbles and Gower area amounting to £132 million (D. Nutt, 1994, personal communication), the tourist industry continues to be vital to the health of the local socioeconomic environment. Set against these pressures, the need to maintain and even enhance the Gower natural environment is equally important. The lack of specific data on carrying capacities, particularly offshore, which has seen the largest increase in recreational users, is still a major problem. The intention to include coastal waters in the next phase of the Gower Management Plan, however, may stimulate such data collection.

Conclusion

Almost 40 years after its designation as the first AONB in England and Wales, the Gower still remains one of the most beautiful stretches of coast in Wales. As such, it provides a useful case-study, from which lessons can be learnt and applied to other stretches of the European coast, particularly the scenic peninsulas of the rocky coasts of northwest Europe, such as Cornwall and Finistère. The early identification of the conservation value of Gower, along with its strict planning controls and management policies, has been essential to sustaining the attractiveness of the landscape of the area. The geography of the peninsula, limiting access to the northern shores, and the strong protective management along stretches of coast owned by the National Trust, have also ensured that the areas of highest nature conservation value have remained more or less unscathed. The objectives of the AONB have therefore been achieved without the policy of no-go areas, preferred by more traditional conservationists. The HC principles have also both been achieved: a wide range of activities, including recreational pursuits, have in fact developed, benefiting from the high-quality environment. Although the HC Forum and Sports Council (1993) are currently advocating the use of resource-capacity zones for recreation planning along HCs in England, it appears that the policy of administrative zones within the development plans for the Gower has been effective so far in restricting recreation and tourism-related development in the most vulnerable, remote zones, such as Rhossili. It could be argued that recreation and tourism have, in fact, brought benefits to the area. The programmes to improve the quality of coastal waters have been spurred on more by the needs of the tourist industry than the demands of the relevant EU directives. In addition, the conservation management of the dune system at Oxwich now relies on visitor trampling to maintain the biodiversity of the habitat.

Evaluating the sustainability of the tourism and recreation industry on the Gower into the next century is fraught with difficulty. The main problem lies in determining the correct balance between tourist development and conservation, not only at site level, but also over the peninsula as a whole. All the habitats, even the sand-dunes and saltmarshes, are seminatural, having evolved over centuries alongside a range of human activities. More environmentally aware recreation and tourism strategies, therefore, can only hope to achieve suboptimal environmental improvements at best. Nevertheless, with careful management Gower will continue to maintain its important conservation status, as well as fulfilling its important role in providing a playground for the urban populations of South Wales and Swansea. However, as recreational pressures offshore increase, so the need for information on the environmental impacts of these will be required so that the waters around Gower can be managed as well.

Pressures from other activities, including residential development on the periphery of the AONB on the urban fringe of Swansea, the impact of offshore aggregate dredging on beach levels, aquatic pollution from the Swansea conurbation and air pollution from the eastern industrialized shores of Swansea Bay, may be greater threats to the Gower environment than recreational pressures. The Gower, lying on the edge of the Swansea Bay urbanized and industrialized embayment, illustrates the problems of many coastal recreational playgrounds. It is the control of all coastal activities, including recreation and tourism, as well as the integration of land- and sea-based management systems, which will, in the end, ensure the long-term survival and sustainability of the high-quality coastal landscapes in Europe such as the Gower. It is to be hoped that the European Commission Council's recent 'Resolution on the future Community policy concerning the European coastal zone' (ECC, 1992) will soon be addressed. This invited the Commission to 'propose for consideration a Community strategy for integrated coastal zone management which will provide a framework for conservation and sustainable use' and recommended that this initiative be incorporated into the European Commission's Fifth Environmental Action Programme.

References

Bridges, E.M., Davies, H.R.J., Goodwin, K. and Hughes, M.R. (1986) The outstanding Gower. *Geographical Magazine* 58(5), 236–245.

CCW (Countryside Council for Wales) (1994) *A Policy Framework for the Coastal and Marine Zone of Wales.* CCW, Cardiff, 12 pp.

Countryside Commission (1991) *Europe's Coastal Crisis: a Cooperative Response. A Report from the European Workshop on Coastal Zone Management, Poole, Dorset, United Kingdom.* Countryside Commission, Cheltenham, 20 pp.

DART (Dartington Amenity Research Trust) (1973) *The Gower Coast: a Report.* Sports Council and Countryside Commission, Totnes, 159 pp.

Davies, A. (1989) Areas of natural beauty: recreation and management. Postgraduate Diploma thesis, University of Wales, Cardiff.

DoE/WO (Department of the Environment/Welsh Office) (1992a) *Planning Policy Guidance: Coastal Planning.* HMSO, London, 21 pp.

DoE/WO (1992b) *Planning Policy Guidance: Tourism.* HMSO, London, 21 pp.

EC (European Commission) (1976) EU Bathing Water Directive, 76/160/EEC. *Official Journal of the European Communities.* EEC, Brussels.

EC (1991) EU Urban Wastewater Directive, 91/271/EEC. *Official Journal of the European Communities.* EEC, Brussels.

EC (1992) EU Species and Habitats Directive, 92/43/EEC. *Official Journal of the European Communities.* EEC, Brussels.

ECC (European Commission Council) (1992) Resolution on the future Community policy concerning the European coastal zone. *Official Journal of the European Communities,* 59/1, 6.3.93. EEC, Brussels.

Gower Countryside Service (1994) *Annual Report of the Planning Department, Gower Countryside Service.* Swansea City Council, Swansea, 9 pp.

HCSCE (House of Commons Select Committee on the Environment) (1992) *Report: Coastal Zone Protection and Planning.* HMSO, London, 380 pp.

Heritage Coast Forum and Sports Council (1993) *Sport and Recreation on Heritage Coasts.* Heritage Coast Forum, Manchester, 46 pp.

Hughes, M. (undated) *Access, Interpretation and Environmental Education: A Plan for Oxwich.* Countryside Council for Wales, Oxwich, 69 pp.

Kay, D. and McDonald, A. (1986) Coastal bathing water quality. *Journal of Shoreline Management* 2(4), 259–284.

Ley 22/1988 de 28 de julio de Costas. Real Decreto 1471/1989 de 1 de diciembre (which establishes the general regulations for the development and implementation of the Ley de Costas).

Morgan, D. (1993) *West Glamorgan Structure Plan Review No. 2, Draft Replacement Plan (Deposit Consultation)*. West Glamorgan County Council, Swansea, 117 pp.

Mullard, J. (1995) Gower: a case study in integrated coastal management initiatives in the UK. In: Healy, M.G. and Doody, J.P. (eds) *Direction in European Coastal Management*. Samara Publishing Limited, Cardigan, pp. 259–284.

NCC (Nature Conservancy Council) (1991) *Nature Conservation and Estuaries in Great Britain*. NCC, Peterborough, 422 pp.

Nelson, C. (1993) The management of coastal conflicts on Gower. MSc thesis, University of Wales, Cardiff.

NRA (National Rivers Authority) (1994) *The Tawe and South Gower Catchment Management Plan: Consultation Report*. NRA Welsh Region, Cardiff, 121 pp.

RSPB (Royal Society for the Protection of Birds) (1993) *A Shore Future: RSPB Vision for the Coast*. RSPB, Sandy, Bedfordshire, 41 pp.

Swansea City Council (1991) *Perceptions of Swansea Study, Stage One: an Investigation into Awareness of and Attitudes towards Swansea and Gower amongst the General Public*. City of Swansea, Marketing and Communications Division, Swansea, 28 pp.

Wilson, D.M. (1990) *Gower Management Plan*. Swansea City Council, Swansea, 92 pp.

Wilson, D.M. (1995) *Swansea Local Plan Review, Written Statement Consultation Draft, April 1995*. Swansea City Council, Swansea, 181 pp.

WTB (Wales Tourist Board) (1994) *Tourism 2000: a Strategy for Wales*. WTB, Cardiff, 51 pp.

4 Landscape and Tourism: *the New National Forest in the United Kingdom*

PHILIP NICHOLLS

Introduction and Background

Forests currently cover 11% of the land surface of the UK. This compares with the Netherlands, a more densely populated country with 9%, but is exceeded by France – 27% – and Italy – 29%. The original aim of the Forestry Commission, established in 1919, was to plant 1,000,000 ha of fast-growing softwoods by the end of the century to act as a strategic reserve in times of war. This 'timber first' policy was remarkably persistent until, in the 1980s according to Mather (1991), the UK entered the phase of the postindustrial forest, a third and final phase preceded by a preindustrial and an industrial stage in evolution. The postindustrial phase is characterized by a more marked shift of emphasis towards the recreational, leisure and tourism aspects of forestry as opposed to the purely timber-producing features. Production of timber, therefore, becomes subordinate to the multiple use of forests, which has now become a central policy objective of the Forestry Commission. However, the delivery of multiple benefits, such as wildlife habitats, fine landscapes, recreation and woodland management, will depend on setting priorities and careful policy-making. In Britain, the conflict of interests in striving to achieve multiple use of forest land has not yet been fully resolved, although the New Forest and the Forest of Dean are classic, well-established examples of multipurpose forestry.

Recent work by Benson and Willis (1993) demonstrates that national forests are places for recreation of many kinds, as well as habitats for a diverse range of wildlife species and an important factor in the character of rural landscapes. Although there are very few relics of natural forest left in Britain, these ancient and seminatural woodlands are a scarce and valuable resource and contain much of the native fauna and flora. Wildlife stewardship, now being actively promoted by the Forestry Commission, attempts to manage Sites of Special Scientific Interest and Forest Nature Reserves, and to promote visitors' enjoyment of wildlife where this is compatible with other conservation objectives (HMSO, 1992). In *Forest Policy for Great Britain* (Forestry Commission, 1991) emphasis on lowland planting, especially broad-leaved species, will help to enhance amenity and wildlife values on land which has been intensively cultivated. Planting in such areas also provides alternative uses for agricultural land producing food crops which are in surplus. The Woodland Grant Scheme, the Farm Woodland Scheme and the Community Forests Initiative will help to afforest more of the lowland landscapes. Furthermore, conservation is now seen as an important objective in attaining sustainability. A report on *Forestry and the Environment* (HMSO, 1993) comments that sustainably managed forests should produce timber on a constantly renewable basis by adopting Worldwide Fund for Nature (WWF) guidelines, against which forest management should be tested: maintaining biodiversity, using renewable resources within their capacity for self-renewal and avoiding pollution and needless consumption. This is a good working definition but sustainable management of forests could also be seen as a concept with an ethical and moral message which 'meets the social, economic, ecological, cultural and spiritual human needs of present generations without compromising the ability of our heirs and successors to meet theirs' (HMSO, 1994).

The UK government's latest strategy on sustainable forestry (HMSO, 1994) sets out eight objectives:

- protecting forest resources;
- enhancing the economic value of forest resources;
- conserving and enhancing biodiversity;
- conserving and enhancing the physical environment;
- developing recreational opportunities;
- conserving and enhancing landscape and cultural heritage;
- promoting appropriate management;
- promoting public understanding and participation.

Although the active promotion of recreation and multiple use of forests has now become enshrined in forest policy, the demand for recreation on the Forestry Commission's estate of 1,000,000 ha has been in place for many years. The first national forest park, Argyll, was created in 1935 and there are now 17 such parks (Forestry Commission, 1988; C. Probert, Edinburgh, 1995, personal communication). The current criteria for the designation of a forest park by the Forest Enterprise are: a minimum area of 1000 ha, high actual or potential use for recreation and a management strategy which reflects multipurpose forestry, with an emphasis on recreation. Fig. 4.1 shows the distribution of these forest parks, National Parks – England and Wales only – and the new National Forest. It will be noted

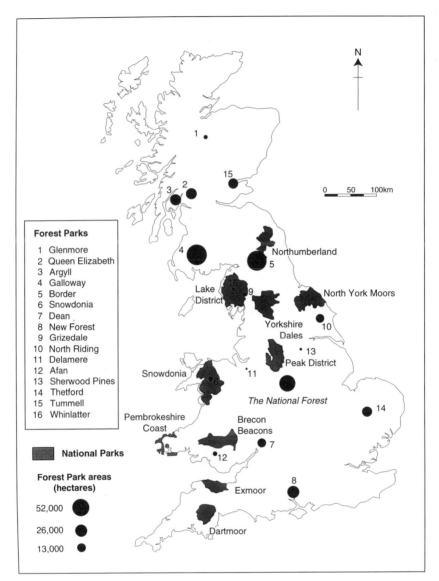

Fig. 4.1. National and forest parks in Great Britain.

that forest parks vary widely in size from the largest, Galloway – 76,100 ha – to the smallest, Delamere – 600 ha. Furthermore, only four – Sherwood Pine, Thetford, Dean and the New Forest – lie in the lowland zone, indicating the relative scarcity of woodland resources here. Those in the highland zone are based on areas with high landscape merit and are particularly favoured by recreationists and tourists on short holiday trips, while those in the lowland zone are designed to attract day visitors and to function as multipurpose community woodlands. The new National Forest, situated in the heart of England, aims to provide an additional user-based recreational resource in an area poorly endowed with community woodland amenities.

In conjunction with the Forestry Commission's informal designation of forest parks for tourism use, the Countryside Acts of 1967 and 1968 gave formal powers to develop recreational potential of forests, especially near large centres of population, and to provide a range of facilities including picnic sites, car parks, trails, visitor centres and camping sites. Pilot studies of recreational demand in such forest areas suggested that there were 10–15 million visits between June and September (Benson and Willis, 1993). It is probable that visits to all woods and forests represent about 10% of all countryside trips.

There are other important influences affecting the demand for forest-based recreation. By the year 2000 there will be a 15% growth in the age-group 30–44 – the most frequent countryside users – plus an increase in the 60-plus age-group. Further, depopulation of inner cities will bring an increase of 3–5 million people into the countryside by the year 2000 (Benson and Willis, 1993). Thus the current 50 million day visits to Forestry Commission woodlands alone each year (HMSO, 1994) indicate that there are likely to be severe pressures on countryside recreation facilities in a few years' time. However, these pressures are by no means uniformly spread. The Forestry Commission estate is relatively remote from centres of population; the highest number of visitors occur in the English lowlands and the lowest in the Welsh and Scottish uplands (Benson and Willis, 1993). This translates into revenues – from recreational activities – of £1 per hectare in the remotest areas and £10–50 per hectare or more in lowland regions (Benson and Willis, 1993). Forests such as Cannock, Sherwood and Epping, with a wide range of recreational pursuits, can generate hundreds of pounds per hectare per year. In such lowland areas these forests may be the only local alternative to the farmed landscape. Local countryside around towns is, therefore, seen as an increasingly important recreational resource, particularly since more than one-third of trips have a round distance of less than 10 miles and 50% less than 20 miles (Benson and Willis, 1993). It should also be noted that National Parks are situated in the highland zone, where recreationists have a greater degree of substitution between sites. Furthermore, access to forests in the private sector is often not encouraged, mainly because there is no financial return and because of conflict with more specialized land uses. However, there is significant use of land for recreation in forests and woods. In 1987 the National Trust recorded 8 million such visits, and 1 million visitors each year converge on Sherwood Forest Country Park and Cannock Chase Country Park (Benson and Willis, 1993).

The preference for and economic potential of forests in lowland areas emphasize the strong spatial demand for and value of recreational use of forests to the south and east of the classical Tees/Exe line. Benson and Willis (1993) calculate that the capitalized value of the Forestry Commission's estate for informal recreation is £883 million, compared with a value of £200 million quoted by the National Audit Office.

Rural areas clearly play a very important role in tourism and leisure activities and, as Shaw and Williams (1994) observed, three-quarters of the population of England visited the countryside at least once in 1990. They go further to say that 'the countryside is socially defined as a premier area of leisure and tourism in modern societies; and visiting the countryside is a socially valued end in its own right' (Shaw and Williams, 1994, pp. 223–224).

In seeking to create a user-based, multipurpose and sustainable forest landscape in the rural heart of the English countryside, the new National Forest will need to adopt and incorporate some of the established recreational planning guidelines currently recommended by the Forestry Commission (1992) to private woodland owners willing to open their estate to visitors. Among their many recommendations, the Forestry Commission emphasize that attractive woodland has several important features. Variety is essential, whether expressed by mixtures of species, different colours of foliage and bark or variation in the age of trees. Contrasts of open space, woodland areas, land and water, shape and form and edges between different features are essential elements in landscape and habitat diversity. Visitor enjoyment is also enhanced by wildlife – birds, mammals, amphibians, reptiles or insects – and plants – shrubs, herbs, mosses, fungi or other forms of plant. Peace and quiet, together with good access and well-signed pathways, further contribute to forest recreation potential. Woodlands, which have a unique ability to absorb the sight and sound of people, are often highly appropriate locations for sports and other, often intrusive, noisy activities.

The new National Forest, like any other forest park, community woodland or recreational area, will need to be carefully designed and sympathetic to a variety of uses. Attractive forests contain a network of open spaces, temporary or permanent, and in designing these the Forestry Commission (1992) suggest using the 'rule of thirds': some areas with two-thirds open glades to one-third trees, other areas with one-third open glades to two-thirds trees. Good forest design ensures that woodlands are in harmony with the character of the wider landscape and takes full account of woodland and wildlife conservation. Although there is little relict ancient woodland in the National Forest, care will be needed in maintaining and enhancing the character of the landscape around the remnants of the ancient forests of Needwood and Charnwood on the northwestern and southeastern edges of the region respectively. Since so much of the contemporary landscape of the

National Forest is currently farmed, it will be more difficult to plan, design and undertake landscape changes *ab initio*.

Despite the difficulties and challenges facing the new National Forest enterprise, the rationale for this venture arises from the need to meet an increasing demand for informal countryside recreation, especially in woods and forests, given the existing and future trends in demography, economy and leisure. The commitment to the multiple use of forest land and the creation of national and community forests by the Countryside Commission and the Forestry Commission are tangible policy expressions of the need to meet this challenge. The concept of a new National Forest based in the Midlands is the vision now put forward by these bodies to satisfy future recreational demand in areas with a low proportion of woodland or forest cover.

National Forests: an International Perspective

Before considering the British National Forest in detail, it is useful to set its development in a wider spatial and temporal context. Early parallels to the creation of a climate in which national forests should be conserved and developed for a wider diversity of uses than just purely timber production can be traced back to Denmark, where for 200 years forests have been governed by state legislation in order to prevent losses caused through exploitation by the navy. Outside Europe, National Forests in the USA were created in 1905 to protect water sources and to provide timber. In 1961 the US multiple-use Sustained Act, with its accent on the management of all resources in forests, was an important precursor to multiple land-use developments in Britain and Europe. A more formalized recreation strategy in the USA was ushered in under the 1976 National Forest Act, which set up a system of developing comprehensive land management plans with full public involvement in each National Forest.

However, in Denmark, a country where two-thirds of the 12% of woodland is privately owned, two agencies – National Forest and Nature Agency – had to be set up in 1974 to integrate production and environmental interests in a multiple-purpose land-use system. In contrast, in Germany, the Federal Forest Act of 1975 provided a national framework in which the states had autonomous powers to develop their own individual forestry policies and programmes in satisfying the aims of wood production and the provision of environmental and leisure benefits. There has been a policy of free public access to all German forests since 1975 and recreation is carefully managed in sensitive areas in consultation with the overseas and local users (HMSO, 1993). These concepts were further developed in the 1989 Forest Act for sustainable forestry, with a special emphasis on woodlands and recreation provision for a population which is 90% urban. The need to provide

recreational outlets for predominantly urban and for ageing populations –
senior citizens will soon be the largest age group in the USA – has become
a paramount objective in forest planning in a number of developed countries.
The British National Forest proposal and vision is clearly a similar response
to an international quest for the maximum sustained use of forestry
resources. The twin goals of increasing woodland area and at the same time
accommodating and absorbing recreational uses are a central feature of this
new approach to forest planning.

The National Forest Proposal

Objectives

The National Forest in Great Britain is both a concept and a vision based
on an imaginative and ambitious environmental project pioneered by the
Countryside Commission in 1987 as part of its case for multipurpose forestry
in the quest to increase the area of woodland and facilities for informal
countryside recreation in lowland England. Five potential sites were suggested in 1987 – Wyre Forest, Sherwood Forest, Arden Forest, Rockingham
Forest and Needwood/Charnwood Forest. All embraced historic woodlands,
farmland and heritage features and were close to large population centres.
The Needwood/Charnwood complex, an area of 502 km² in the English Midlands, was chosen in preference to the other four, because of its low amount
of woodland cover, numerous derelict land areas based on mineral workings,
good communications and access to large population centres, massive support from the local landowners and population and the prospect of linking
two separate blocks of ancient forest in a network of newly created
woodlands.

The aim of this forest proposal is to create and demonstrate a truly
sustainable and multipurpose forest that meets the following environmental
and economic objectives:

- to enhance and create a diverse landscape and wildlife habitat;
- to create a major recreation and tourism resource;
- to provide an alternative productive use for agricultural land in a manner
  that meets environmental objectives;
- to contribute to the national timber supply;
- to stimulate economic enterprise and create jobs;
- to contribute to wider environmental objectives such as a reduction in
  carbon dioxide in the atmosphere (Countryside Commission, 1993).

Although the concept of the National Forest is not necessarily new, the
intention is to model it on the New Forest, Hampshire, which has an attractive blend of historic, scenic, ecological and environmental characteristics.

However, the New Forest represents an intact survival of a Norman land-scape, which has been influenced by the custom of regulated grazing rights, human influence and browsing practices. The New Forest is now classed as a World Heritage Site because of its enormous ecological and historic importance. The National Forest as proposed may strive to emulate aspects of the general character of the New Forest but it is a very different environ-ment, dominated by farming and industry and with very little of the heath-land and ancient and ornamental woodland which characterize the New Forest. The National Forest will, therefore, be a purposeful development and control of land uses which must accommodate farmers' interests and the need to provide and satisfy considerable recreational demand from the sur-rounding region. In order to understand how these goals might be realized, it is necessary to discuss the National Forest's main characteristics, land uses, tourism potential, agricultural and conservation features in some detail as a basis for assessing the degree to which goals of sustainability might be attained.

Characteristics

The area occupied by the proposed National Forest lies between the cities of Birmingham, Derby and Leicester and stretches across the counties of Staffordshire, Derbyshire and Leicestershire (Fig. 4.2). The Forest consists of three major sections, the ancient forest remnants of Needwood in the west,

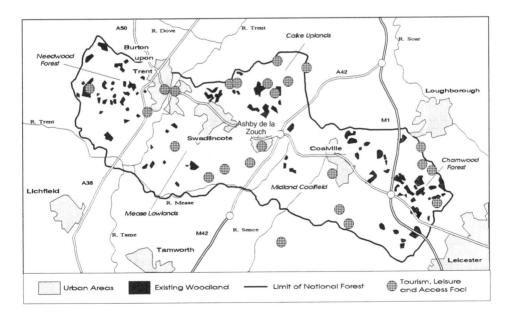

Fig. 4.2. The National Forest.

Charnwood in the east and a substantial portion of the Midland coalfield in the middle. The site itself is not physically very distinctive or outstanding compared with, for example, the New Forest or Forest of Dean. The relief is gentle and undulating. In the west, the Needwood Plateau rises to 135 m and is capped by extensive deposits of heavy boulder clay dissected by many streams and marked by prominent river bluffs overlooking the River Dove in the north and the River Trent to the southeast. The Trent system runs northeast across glacial sands and gravels and is edged by flat river terraces. The River Mease lowlands to the east and southeast of the Trent, developed on Mercia mudstone, rise to 100 m towards the coalfield plateau, which comprises coal measures and Keuper sandstones. Carboniferous limestone formations and millstone grit outcrop on the northern fringe, where the land-forms are more undulating, at an altitude of approximately 150 m. Eastwards, the land rises sharply to the Charnwood uplands, which mark the highest part of the area, at 278 m. Geologically this is the oldest area and consists of Precambrian rocks exposed through the surrounding Keuper marls and mudstones. Here the landscape has a more rugged, rocky appearance.

These physical components of the landscape are reflected in the historical geography. In the west lies the former Royal Chase of Needwood, which remained under Forest Laws until disafforestation – the legal extinguishing of common rights preceding enclosure – in 1804 (Nicholls, 1973). Here the cultural landscape is distinguished by peripheral forest-side settlement as a result of Anglo-Saxon and medieval encroachments together with the remains of several deer parks and irregularly patterned fields. The landscape of the forest core, as defined by eighteenth-century surveys and the 1804 enclosure proposals, comprises a network of contrastingly square, regular fields, scattered farms, remarkably straight roads, pockets of woodland which survived enclosure and a number of wooded parklands where the former lodges stood. In contrast, Charnwood Forest in the east of the National Forest site did not survive for nearly as long as Needwood and functioned as an extensive common pasture and waste dominated by heath vegetation. Most of the Charnwood woodlands were felled to provide charcoal for the region's many iron forges. The coalfield section has a very different landscape, as a result of mining activity since the thirteenth century, but more particularly in the nineteenth and twentieth centuries. The historic settlement pattern was closely related to woodland shrinkage and agricultural and industrial developments. Nucleated villages, scattered farms and small mining communities are very characteristic. Distinct building styles can be recognized, ranging from red brick and timber in Needwood, sandstone vernacular around Melbourne and brick terraced cottages in the coalfield, to traditional granite, stone and slate buildings in Charnwood.

Table 4.1 gives a summary of the main land uses, land quality, woodland and farmland in the proposed National Forest. The landscape character of the proposed National Forest is clearly dominated by farmland – 76% – and, even though 54% is classed as Grade 3, the creation of a multipurpose forest environment will depend ultimately on whether farmers will be prepared to invest in forestry, allow access and embark on a considerable switch in land uses. Safeguarding the 10% of Grade 2 land should also be a priority.

Table 4.1. National Forest: some characteristics, land uses and land quality (from Countryside Commission, 1993).

Characteristic	Total	%
Population	187,000	
Total surface area	502 km²	
Land uses		
Mineral working – sand, gravels, quarries	1000 ha	2
Open-cast coal working	1000 ha	2
Derelict land	500 ha	1
Woodland – 36% broad-leaved, 9% conifers, 55% mixed	3010 ha	6
Sites of ecological importance – 50% Charnwood	4700 ha	9
Sites of Special Scientific Interest (SSSIs) – 26 sites	1800 ha	4
Farmland – including woodland, sites of ecological importance and SSSIs	40,000 ha	76
*Land quality**		
Grade 2		10
Grade 3		54
Grade 4		10
Non-agricultural land – golf courses, urban land, mineral workings, allotments		26

* According to MAFF (1966) category designation for land-use planning, based on soil, relief and climatic conditions:
Grade 1: Land with very minor or no physical limitations to agricultural use. Deep, well-drained, sandy loams, silt or peat lying on level sites or gentle slopes, all of which are easily cultivated.
Grade 2: Land with some minor limitations, e.g. in soil texture, depth or drainage. A wide range of agricultural or horticultural crops can be grown.
Grade 3: Land with moderate limitations – limiting factors include altitude, slopes between 1:8 and 1:5, and annual rainfall over 1014 mm. Grass and cereals are principal crops.
Grade 4: Land with severe limitations – land over 183 m, annual rainfall over 1270 mm – and low productivity. The main products are oats and barley.
Grade 5: Land with very severe limitations, e.g. very steep slopes, poor drainage, land over 305 m, annual rainfall over 1524 mm. The land is used for rough grazing.

Development strategies

The main strategies proposed by the Countryside Commission for the long-term creation of the National Forest are part of the overall vision to satisfy a number of competing claims in a sustainable context. Whether or not these goals and aims can be achieved and sustained will depend on many factors, some known, some unknown. However, in order to evaluate the degree of conflict between land uses and objectives, it is necessary to discuss briefly the main options being pursued.

Firstly, the forest strategy is focused on forest planting, sport, recreation, tourism, agriculture, nature conservation and historic heritage, built development and derelict land. The new forest resource attempts, in a sustainable way, to satisfy a number of competing objectives.

Secondly, forestry, a key element in the National Forest, has been modelled on the New Forest in terms of land-use mix in an attempt to blend historic woodlands, commercial forestry enterprises and recreation amenity woodlands. Multipurpose woodlands would cover around one-third of the area, and 70% should be established in 10 years. Types of planting suggested are plantation and farm woodlands, especially in the Mease lowlands, rural coalfield and Calke uplands; recreation and amenity woodlands; networks of wooded corridors; and interlocking patterns of forest, open ground and wetland. Country and forest parks, as well as community woodland planting, would help to benefit urban fringe landscapes. Energy forestry, especially willows and poplar, would be suited to the moist, fertile areas of the Trent Valley. Game woodlands are suggested for the wooded parklands and enclosed farmland landscapes of Needwood, Calke uplands, Midland coalfield and Charnwood. Agroforestry, an intimate mixture of trees with farm crops/animals for rotations of 20–45 years would be especially suitable for arable landowners. Urban forestry, based on local initiatives for individual towns and villages, could improve the appearance of many settlements.

Thirdly, one of the central aims of developing a National Forest is to contribute to a successful recipe for meeting a growing demand for participatory enjoyment in the countryside. The Forest would provide a recreational resource for the local community of 187,000, a population of 10 million within a 90-minute drive for day-trips and 29 million within $2\frac{1}{2}$ hours' drive for short breaks. It is estimated that the Forest could attract at its peak nearly 6 million visits per year, 90% of which would be day-trippers (Benson and Willis, 1993).

A recent questionnaire survey (Countryside Commission, 1993) targeted at nearly 20,000 people showed that, while 99% supported the National Forest and 80% agreed that the area should be developed for leisure, sport, recreation and tourism, there were fears that these objectives might be difficult to fulfil in practice. For example, the Ramblers' Association was unhappy about the freedom of access to roam. Permissive access is not seen as sufficient, and there is pressure to create several thousands of hectares of environmentally benign, multipurpose woodland, offering free access. The area is very accessible via the M1 and M42 motorways, major trunk roads and inter-city rail links which tend to run north–south. However, east–west and internal routes are often narrow and tortuous and could be severely strained under visitor pressure. Trails, bridleways and access routes are relatively poorly developed and careful planning will be needed to provide adequate visitor access.

It is hoped to develop sport and leisure activities through the reuse of surplus agricultural land and reclamation of derelict land, accommodating canoeing on chains of flooded, disused gravel workings and using woodland areas for archery, war games and shooting. Golf courses are also planned but these need to integrate with the landscape, especially where important historic features, such as deserted settlement sites, ridge and furrow, ancient hedgerows and specimen trees exist. Visitor attractions, particularly historic properties, museums and galleries, wildlife attractions and gardens are poorly represented in the National Forest. However, there are some attractions which already attract large numbers of visitors. For example, the Country Parks – including Bradgate Park and Beacon Hill, both highly visited – received 1,500,000 visitors in 1993 and the National Trust-owned Calke Abbey, Bass Museum at Burton, Byrkley Garden Centre and Staunton Harold Craft Centre received 611,000 (Table 4.2). Facilities for user-based sports also exist and can be further developed. These include angling, golf,

Table 4.2. National Forest: recreation, leisure and tourism (based on Countryside Commission, 1993, and National Trust, 1995).

Activities and attractions	No. of facilities	No. of visitors 1993
Resource-based activities		
Public access areas – Country Parks, woodlands managed by Woodland Trust	23	
Water areas	8	
Walking routes/trails	24	
Tourist attractions		
Historic properties	12	
Calke Abbey		106,000
Bass Museum		80,000
Country parks, including Bradgate Park and Beacon Hill		1,500,000
Byrkley Garden Centre		325,000
Staunton Harold Craft Centre		100,000
Total		2,111,000

sailing and windsurfing, canoeing and rowing, riding, clay pigeon shooting and motor sports, together with leisure and recreation centres. The existing leisure and tourism facilities in the National Forest, shown in Table 4.2, suggest that a number of additional user-based recreation outlets will have to be developed. The location of these developments is shown on Fig. 4.1. Nevertheless, it is possible that the National Forest has neither a sufficient mix of visitor attractions nor an adequate provision of hotel and other accommodation to absorb a projected 6 million visitors per year.

Fourthly, the National Forest will embrace an area of active agriculture, which is a vital income source to farmers, tenants and the region overall. The degree to which much of the land can be turned over to forestry and recreational enterprises will depend on farmers' attitudes and incomes and the wider issues of the EU Common Agricultural Policy (CAP) and Set-Aside Policy. A major difficulty in developing forestry activities will be caused by land tenure. It is estimated that 41% of agricultural land in the Forest is tenanted (Countryside Commission, 1993) and that this will inhibit the progress of forest planting schemes. However, schemes to enhance farmers' incomes from less traditional activities include farm diversification and rural enterprise through farm tourism. This could be achieved through the provision of serviced and self-catering farm-based accommodation, using barns and outbuildings, angling lakes, clay pigeon shooting, golf, picnic sites and farm woodland walks and the development of farm parks.

In addition to these schemes, farmers could develop permissive riding routes on their land and toll rides. Additional income-generating outlets for farmers in the National Forest involve tearooms, restaurants, pick-your-own, farm shops and speciality food outlets. The Forest strategy could generate new rural industries based on timber and wood products – sawmills, craft industries, charcoal, garden chips and fuel wood. Short-rotation coppice on a 3–5-year cycle could help to nurture an interest in longer-term forestry.

Fifth, the National Forest strategy plans to improve nature conservation through a restructuring of ancient woodland, especially Charnwood, and, to retain ecological diversity, woodlands should be managed in a multipurpose way with wide, open, grassy rides edged with shrubs to provide habitats for pheasants and butterflies and to act as dry extraction routes for timber. The plan envisages up to 20% of woodlands being set aside as open habitats. Wildlife corridors, which include hedgerows, roadside verges, derelict railway lines, watercourses and forest trails will be conserved within the matrix of larger forestry and multiuse schemes.

To change the existing land uses so that up to one-third of the area is planted with a mixture of conifers and broad-leaved species will require considerable cooperation by farmers and landowners. At present the Farm Woodland Scheme, set up in 1988 with the aim of diverting land from agricultural production to help reduce farm surpluses, to develop recreational potential and to contribute to farm incomes through timber production, has had a disappointing uptake. Ilbery and Kidd (1992) show that in England only 12,000 ha per year were converted between 1988 and 1991, well below the anticipated 36,000 ha per year; this is despite the fact that grants of up to £505 and £1,375 per ha are allocated through the Forestry Commission for plots of 1–2.9 ha. The scheme is targeted primarily at areas of arable land and improved grassland, and a farmer must plant a minimum of 3 ha but no more than 40 ha to qualify for the grants; each block of woodland has to be at least 1 ha. The disappointing rate of uptake for this scheme and the Ministry of Agriculture, Fisheries and Food Set-Aside Woodland Options scheme appears to be related to existing patterns of farm woodland, the lack of targeting success in the core arable areas of eastern England and the fact that current Set-Aside policies require farmers to rotate the 15–20% of this arable area taken out of production – clearly quite impractical for forestry purposes. As far as the National Forest is concerned, dramatic landscape changes and uses are unlikely to be achieved unless considerable government funding and policy changes are forthcoming.

Nevertheless, the concept of the National Forest is now firmly embedded in policy-making and should be seen as part of a much wider spectrum of woodland and forest recreational facilities in Britain as well as in the wider, international context.

Conclusion

The creation of a new National Forest in Britain aims to solve a number of often conflicting goals within a sustainable framework. In attempting to reconcile these aims there will, however, be a number of problems. Firstly, forestry and woodland could occupy 166 km^2 (33% of the present area), but this would be mostly at the expense of farmland which is privately owned or tenanted. Thus, existing land-tenure structures, uncertainties over the CAP, the slow uptake of Farm Woodland Schemes and other tree-planting incentives suggest that it will be very difficult to increase the woodland area and improve landscape appearance without financial or other support. The availability of finance, together with cooperation by landowners, is essential to the realization of this forest scheme.

Secondly, and most importantly, the provision of recreation, tourism and leisure facilities within the newly created forest environment is also fraught with problems. Among these are east–west access across the Forest and on to private land, the availability of adequate visitor attractions to accommodate anticipated demand and the provision of user-orientated recreational facilities on new sites. The Ramblers' Association feels that many Midland landowners may not be prepared to accept an obligation to provide reasonable public access, with the result that the National Forest could become a fenced-off private amenity for local landowners (HMSO, 1993).

Thirdly, paramount to the success of the National Forest is the creation of a more wooded landscape interspersed with farmland pockets, wildlife, heritage, conservation and other designated sites to which the public has easy access. The aim is to accommodate recreation and tourism in a sustainably managed wooded landscape. Unlike the Peak District National Park, the Lake District or the New Forest, the National Forest has no intrinsically spectacular, scenic or special qualities. Its success will depend primarily on opening up the area to walkers, joggers, pony-trekkers, anglers, etc., as a means of satisfying an increasing demand for leisure and recreation facilities in this urbanized area of the Midlands. The National Forest would provide a welcome and much needed alternative recreation and tourism venue to the already highly congested and overvisited National and forest parks and would, therefore, indirectly help to sustain landscape quality in other parts of the country.

Fourthly, the Forest will have a major impact on the region's economy. Sustainable woodland and tourism management will enhance the economic prospects of the region and help to revive agricultural prospects. However, the dramatically changed economic circumstances of the 1990s, coupled with government emphasis on cost accountability and reducing public expenditure and, above all, the continued uncertainty over the long-term future of the Forestry Commission, suggest that the Forest may have a slow and unpredictable development. Major countryside changes have been taking place over the last 10 years. The National Forest is a reflection of the change towards a postproductionist, less intensive, agricultural landscape, now more closely focused on maintaining and improving ecological diversity and public access.

Finally, forestry operations and tree-planting schemes are intrinsically long-term and require consistent, forward-looking, stable policies and careful decision-making. Sustainable management, with its emphasis on satisfying current needs without jeopardizing future provision, would seem to be the best way of achieving a truly multipurpose National Forest.

References

Benson, J.F. and Willis, K.G. (1993) Implications of recreation demand for forest expansion in Great Britain. *Regional Studies* 27(1), 29–39.

Countryside Commission (1993) *The National Forest Strategy.* Countryside Commission, Cheltenham, 88 pp.

Forestry Commission (1988) *Forest Parks.* Forestry Commission, Edinburgh, 2 pp.

Forestry Commission (1991) *Forestry Policy for Great Britain.* Forestry Commission, Edinburgh, 45 pp.

Forestry Commission (1992) *Agenda for Forest Enterprise.* Forestry Commission, Edinburgh, pp. 19–38.

HMSO (Her Majesty's Stationery Office) (1992) *Forest Recreation Guidelines*. HMSO, London, 36 pp.

HMSO (1993) *Forestry and the Environment*, 2 vols. HMSO, London, 180 pp.

HMSO (1994) *Sustainable Forestry – the UK Programme*. HMSO, London, Cm 2429, pp. 24–35.

Ilbery, B. and Kidd, J. (1992) Adoption of the Farm Woodland Scheme in England. *Geography* 77(4), 363–367.

MAFF (Ministry of Agriculture, Fisheries and Food) (1966) *Agricultural Land Classification of England and Wales*. Technical Report 11, MAFF, London, pp. 18–24.

Mather, A.S. (1991) Pressures on British forest policy: prologue to the post-industrial forest? *Area* 23(3), 245–253.

National Trust (1995) *Facts and Figures Compendium*. National Trust, London, 30 pp.

Nicholls, P.H. (1973) On the evolution of a forest landscape. *Transactions of the Institute of British Geographers* 56, 57–76.

Shaw, G. and Williams, A.W. (1994) *Critical Issues in Tourism: A Geographical Perspective*. Basil Blackwell, Oxford, 280 pp.

5 Tourism and Local Awareness: *Costa Brava, Spain*

Arthur Morris

Introduction

The main concern of this chapter is the evolving awareness of environment and related social and economic issues, over the course of a period of tourist-related development. In the literature on tourism development, frequent reference is made to the concept of a cycle, starting with discovery, leading to limited specialized development of an élite market followed by local firms increasing the market to a respectable size and expanding the range of interest to a broader group of customers – see, for example, Miossec (1976) in Pearce (1995). In later stages international tourism may take over the industry, degrade the resource and then abandon the area, leaving in its wake a crisis region whose economic base has been lost. The concept is one of an economic activity which is not sustainable, and which moves on from place to place consuming fixed local resources. Miossec's model includes local changes of view in parallel with changes in the industry. One theme of this chapter is that these changes may be quite autonomous of the tourist industry's own change. Massive changes in social structure, urbanization and national and regional politics may be expected to produce their own effects amongst the population of the receiving area. Separate models of attitudes in the destination areas have been produced, such as Doxey's Irridex model (1975), or that of Butler (1975), both cited in Murphy (1985). These models, however, look at the tourist cycle in isolation, whereas it is proposed here that a view on the contemporaneous evolution of society generally is needed.

There are no diachronic studies that fully document these changes, and their rate of action is so gradual that field-study methods, such as the use of questionnaires, are not practicable. But much information may be gleaned from secondary data of various kinds, and this is what has been used in the present study. Like the tourist cycle, a set of phases may be posited, though these do not follow with any precision those of the tourist industry itself. Cultural identity and the gap between the receiver society and the incoming tourist society condition the change. Most of the information has been gathered from the region of the Costa Brava in Spain, which is an appropriate study area because of its long history of tourist development (Fig. 5.1). Given the intensive development of the region for mass tourism by charter companies, the case may be thought extreme, but it is not unrepresentative of many Mediterranean areas.

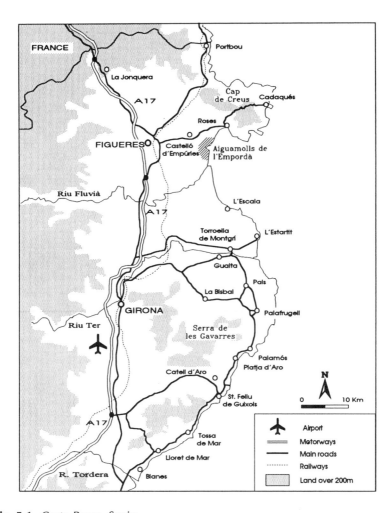

Fig. 5.1. Costa Brava, Spain.

Phases of Spatial Expansion of Tourism and of Reaction from the Local Population

In this section attention focuses on the geographical or spatial aspects of the growth, examining both action and reaction. In the first phase, tourism was punctiform, concentrated at just a few separate points on the coast of the Costa Brava, where first family hotels and then massive tourism, from about 1960, were developed. Concentration in the first instance was due to the tiny total demand, and later because mass tourism was based on large groups brought in by air to big hotels providing cheap holidays by achieving considerable economies of scale. In the 1970s, the demand for new tourist space switched to second homes and camping grounds, plus recreation facilities, such as swimming pools, go-kart courses and marinas, and the pattern became linear, strip development along the coast. This was obviously more space-consuming, though still quite concentrated. In a third spatial phase, the demand became quite diffuse, as tourists sought further diversions outside the coastal area, became more mobile by using buses or private cars, and moved inland as the coastal area became less desirable. In this latter phase, the whole countryside was a potential target for the tourist. Throughout this process there appears to be an inevitable dynamic, whereby the tourist is constantly seeking a new experience (Urry, 1990) and this restlessness can be partly satisfied by moving into new places within the tourist region.

Phase I: naïve reception

Alongside the changes just described, and partly in reaction to them, there are changing local attitudes to the tourism activity. In the first phase, that of concentrated development, the attitude to tourism is likely to be positive and receptive. On the Costa Brava this was certainly the case; a declining cork industry and other small industries producing brandy and ceramics had left this region with some accumulated wealth in the mid-twentieth century and considerable entrepreneurship, which was early invested in small family hotels in the fishing villages along the coast, especially in the 1950s.

It is worth noting the spontaneous character of this early evolution of both demand for and supply of tourism. In some analyses (Cals, 1974, especially Chapter 3; Tamames, 1977, Chapter 9), the tourist industry's development in Spain, including the Costa Brava, is attributed very largely to the promotion of central government. Madrid controlled prices, provided lines of credit, set up public enterprises and built infrastructure. But for the Costa Brava there was very little state help for the development of the industry before 1973 (Cals, 1982, pp. 54–55). Finance was unimportant, and investments in infrastructure or public services relied on the provincial and municipal governments and chambers of commerce. Development depended on local enterprise and private investments in small family hotels. Thus a positive attitude was needed and was found amongst these people. The example of Girona airport is illustrative. This airport had been petitioned for from central government throughout the 1950s, but was never granted, so that finally the land was donated by the province of Girona and 60 million pesetas were contributed by the city council and the private sector in order to finish the airport, as central funding came too late and was insufficient in quantity.

Other indicators of positive local support for the industry in the 1950s and 1960s are the local plans. Each municipality, under Spanish law, is required to produce a *Plan General de Ordenación Urbana*, a physical plan covering land uses but not economic and social development, comparable to British local plans, for its territory. In the absence of higher-level provincial or regional plans, these plans became the main guide to land use in Francoist Spain, especially after the Law of 12 May 1956, which was used until reformed in 1975 (Jiménez Luna, 1983). This situation, plus the lack of skilled planners and the strength of the town council in controlling planning, meant a great variety of outcomes in the field. Thus, for example, the municipality of Palafrugell, with a conservative mayor and council, had a very modest development of tourism, leaving large areas as farmland or forest. In contrast, Lloret de Mar, at the southern end of the Costa Brava, became a mass destination for tourist companies. Its town council was dominated in the 1970s by a mayor who actively promoted development, whether by the charter companies or through the plan allotting land to second-home construction, and a huge infrastructure capacity was created, including a water-supply sufficient for a population of 250,000.

Study of the *Plan General* documents from the 1960s gives a good idea of the reception they were given. At Lloret de Mar and similarly at Castelló d'Empúries – in contrast to later plans for this town – the massive development proposed under the plan, when it was brought to public exhibition stage as required, was opposed by a tiny minority. Objections were on the grounds of loss of land to roads or infrastructural developments of other kinds, preventing individuals from realizing their own private and small-scale developments, rather than any concern for the proposed scale of development, the new hotels or other industries brought in, the new demands on water, the degradation of the visual landscape or any kind of visual pollution (Morris *et al.*, 1989). Cases of moderation and planning control, such as that at Palafrugell, were the exception. From all this planning trend, there is a strong inference that the reception given generally to tourism in the Costa Brava was welcoming and positive.

Phase II: limited awareness

The spatial second phase, following the outline indicated above, is the linear or strip development of the coast in the late 1960s and early 1970s. This phase elicited a response which came from social groups with a high level

of education and awareness of ecological, historical and cultural values, and
these groups found much to criticize in the expanding industry.

There had been some nature conservation in Spain from early in this
century and before, so that it is not possible to speak of awareness of environmental concerns moving from an initial situation of total apathy. But the
kind of conservation measures adopted in Spain indicates very clearly the
limited interests at work in the country until very recently, at least at the
level of effective action. Specifically, the National Parks (Fig. 5.2) show
clearly that areas chosen for protection would be those with wild nature,
such as the High Pyrenees, the Coto de Doñana or the volcanic peaks of the
Canary Islands (Morris, 1992). The only real exception is the Covadonga
Park, with its historical connections. No parks were created to conserve
traditional rural landscapes. From the 1960s another park designation was
used in Catalonia, the Nature Park – *Parc Natural* – a regional park which
might have settlements in it, but even these were mostly in the high mountains or set up as green-belt areas around Barcelona. In the 1990s other
designations of protected areas have emerged in Catalonia, but these run on

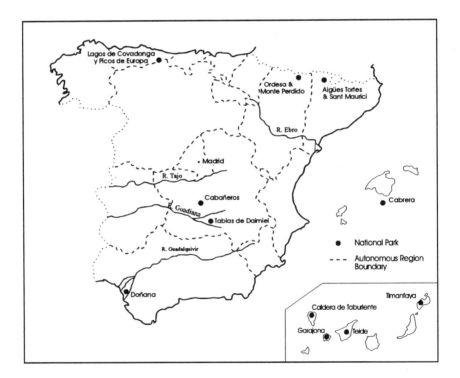

Fig. 5.2. National Parks in Spain.

into the Phase III period described below. In Spain, conservation movements – the word conservation was not popular and they were usually called
ecological movements – became socially and politically salient in the 1970s
(Recio, 1992). Their concern was for scientific interests such as the survival
of individual species and the habitats they might need in order to survive.
Because of the increasing impact of tourism on the coast, these movements
came into conflict with the industry, though only in reference to specific
sites of high ecological value.

On the Costa Brava, a first big success was with the defence of the
Aiguamolls – wetlands in Catalan – of the plain of Empordà (Morris *et al.*,
1989). Here, in the municipality of Castelló d'Empúries, a large and, in terms
of the planning regulations, illegal urban development, Empuriabrava, had
grown up since the late 1960s (Fig. 5.3). A further urban project was

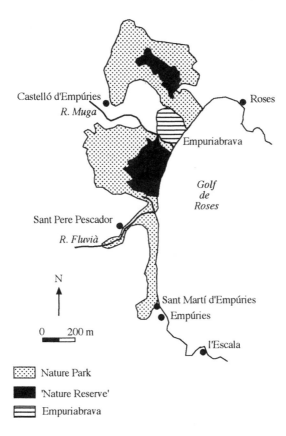

Fig. 5.3. Land use on the Ampurdà coastal lowlands (from Saurí Pujol *et al.*,
1996).

planned by the Port Llevant company, to extend the planned urban residential estate – in Spain called an *urbanización* – on to some 870 ha of wetlands to the south (Planas i Roig, 1987, Chapter 29, pp. 78–79). Inquiries and debates over this plan, which was opposed mostly by conservationists because of the high value of the wetlands for birds migrating between North Africa and Europe, went on between 1976 and 1982. As a result, a small and narrow set of interests was able to check a large-scale economic interest, which would have further developed tourism in the region, and, instead of the new marina and residential development, a Nature Park was established in 1983 and, within it, a Nature Reserve, where access and economic activity are banned (Fig. 5.2). Following this first success, other proposals for development of the various wetlands along the coast have been turned down, mostly without getting to the stage of any public debate. For the purposes of the present study, the significant matter is that this movement was not a gradual stirring of public opinion against tourism in general, as posited in the models cited in the Introduction. Instead, it was a sophisticated defence of ecological interests by a small number of people, involving one specific area on the coast. Only at later stages is more generalized reaction observable.

Phase III: countryside conservation

From the 1980s it becomes apparent that tourism of different kinds may have an impact, not just on the coastal strip, but also over a band of territory of some width inland from the coast including ordinary rural landscapes as opposed to those with high nature conservation value, rural settlements and many places of historic interest. Concern over the negative effects of tourism becomes wider-based in terms of the proportion of local population involved, and wider too in terms of the kinds of concerns which are aired. New concerns are not to conserve nature alone, but to conserve ways of life, landscape in general and places and areas of historic significance.

Many issues were first enunciated and brought together in the *Debat Costa Brava* (COCIG, 1978), a publication resulting from a conference held in the autumn of 1976. This was not a standard academic or business conference, but a meeting of all those interested in the Costa Brava, including business and academic voices but also those of local authorities and others concerned for the region. This meeting marks a crossroads between Phases II and III. It involved both specialists with scientific interests, corresponding to the earlier phases, and some wider interests; it also concerned itself with both the coastal development and some more general issues. Besides the conservation interests, there was a discussion of the economic changes wrought by tourism, such as rapid growth of the service sector at the expense of farming and fishing; the breakup of traditional society through new opportunities and exposure to West European societies and social standards; and the radical alterations to both the nature of the physical land-scape and human interaction with this landscape, including the loss of access to territory, both the coastal strip and the former open lands around the fishing villages, through the growth of the *urbanizaciones*. Much of the new critique concerned the spatially extensive character of the new tourism. Rather than the old, concentrated activity, which affected only single points like Lloret de Mar and Tossa de Mar, bringing in tourists by air and flying them out in a sanitary operation which left the landscape unscathed, 1970s tourism was now seen to have involved second homes and the growth of unsightly *urbanizaciones* – estates whose roads might be unpaved and whose infrastructural services were often incomplete, and without any architec-tonic relation to the old villages. It also involved such space-consuming land uses as racetracks, swimming-pools, camping grounds and many roads to cater for tourists arriving by car or bus rather than plane. Juli Esteban (Esteban Noguera, 1978, p. 21) refers to all of these as parasitic develop-ments and as privatization of the territory, in the sense that old farmlands around the villages, privately owned but allowing open access to local people in bygone times, had now become inaccessible. Other criticisms were of the reorganization of space, away from the old service centres and fishing vil-lages. These had been set carefully through the historical processes of settle-ment by peasant farmers and fishermen, in relation to the resources of the countryside, to needs for defence or to water-supplies. Now the trend was towards a massive strip development along much of the coast, which was fundamentally indifferent to topography, water or soil resources.

The culmination of concerns for ordinary countryside might be regarded as the Plan of Areas of Natural Interest – *Pla d'Espais d'Interès Natural*, or PEIN (Generalitat de Catalunya, 1992) – which from 1992 designated areas as of landscape value for protection. In the case of the Gavarres (*vide infra*), one of the PEIN areas, the basis for the designation is for its geology and physiography, giving rise to a distinctive vegetation and internal diversity. It should be noted, however, that the PEIN represents only a 'negative' desig-nation, in contrast to the positive endowment with resources and plans that are possible for parks. PEINs are confined to negative measures, restrictions on urban and rural planning regimes, on top of those already in place within the municipal General Plan.

There is not room here for the analysis of changes in tourist demand which occurred in the region. They are comparable to those elsewhere in Spain, because the demand effects operate on a Europe-wide scale. Transport evolution and better access to the country allow movement from charter planes to cars and coaches, implying a wider spread of tourist impact. Con-stant familiarization with individual regions means a move to extend out everywhere to new places or attractions, a further extensification of demands. Sectorally narrow demands for sun and seaside evolve towards demands for a variety of more active recreation experiences, extending tourism into camping and visits to heritage sites. In the Costa Brava, there is the additional feature of rapid evolution from one form, intensive, to another, extensive. This period is characterized by an active seeking for new experiences, new places and new positions, which involves the tourist in consuming values that are not then available to others. Like the suburban house on the edge of town, the tourist seeks a second home or a hill track which has not yet been found by others. What may be termed a positional value (Urry, 1990, pp. 42–44) – the value of owning or of having experienced something different from others – drives this kind of modern tourism, although, because of the seeking by many new tourists to experience something authentic of the region visited, there may be a less visible impact than traditional tourism. It must not be inferred that the relations of tourism's spatial spread and the rise of public feeling are closely synchronized; growing and broadening criticism of the nature of tourism development was indeed related to the actual growth of the industry, but also, as will be commented on again at the end of the chapter, to the more liberal atmosphere developing in post-Franco Spain and to social and demographic change in Catalonia and Spain.

Case-study: the Gavarres

One well-known case of a rising concern for countryside has been that of the Gavarres, a hill chain between Girona and the coast (Fig. 5.4). Rather than species or habitat conservation, this case represents a concern for landscape conservation. It is a forested hill mass with a long history of active management for productive purposes. From the late eighteenth century, the main product has been cork from the cork oak, which grows well here. If cork extraction is to be maintained, the whole forest must be actively managed, including such tasks as the upkeep of roadways on to the properties, opening up the forest to remove competing shrubs and trees around the young cork oaks and the extraction of the resulting dead brushwood and prunings which might provoke fires. The cork industry traditionally involved all the small towns and villages around the hill mass, as the rural economy was typically one of farmers who owned some arable land, some pasture on the slopes and in the hills a piece of forest used to extract cork, firewood and other products. Many isolated farmhouses are scattered in and around the forest area, testimonial to the intensive former use of the forest, though some are now abandoned.

Conservation of this area thus implies the active management of a worked landscape. The central problem with putting such management into place is that the cork industry is in decline and has been for many years. In Catalonia, through the competition of tourism and other non-rural employment oppportunities, rural wages have risen rapidly in the last three decades.

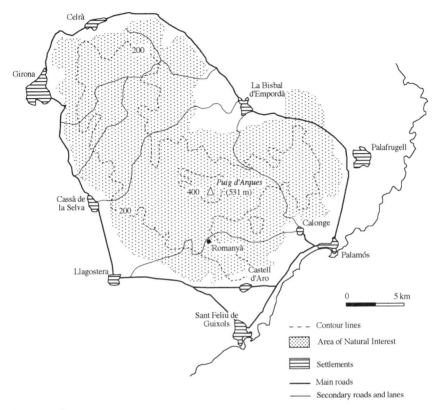

Fig. 5.4. The Gavarres.

Current wage levels make cork extraction and forest maintenance, some of it requiring special skills and using contracted labour, too expensive to contemplate for small farmers. There is also strong competition in cork production from the main producing areas of Extremadura and Andalusia in Spain, where wages are lower, and from southern Portugal. Poor maintenance in the Girona forests has meant that the cork of the few producers still operating is of lower quality, with irregularities and disease and thinner and coarser than before the advent of tourism, when the area was renowned for its cork.

If one side of the problem for the Gavarres is its declining economy, the other side is its ecological richness. This hill mass represents a unique habitat, the last evergreen oak forest in a reasonable state of preservation adjoining the western Mediterranean. Its acidic soils, contrasting with the limestones most typical of hill masses around these coasts, favour the cork oak, as does the relatively high rainfall. In contrast with the cork oak forests of other parts of Spain, much of the forest has become, with the abandonment of regular cork extraction and little competition with livestock, a diverse and rich forest, not the *dehesa* type of park landscape of Extremadura and Andalusia, in which the trees are relicts and eventually give way to grass (Ortega Hernández-Agero, 1989, pp. 97–99). In the Gavarres, in field studies including 5 m square quadrats of forest land, up to 35 woody species were encountered in single quadrats, a richness approaching that of undisturbed Mediterranean evergreen oak forest (AAGA, 1992). This richness of flora and vegetation is also attributable to the location of the wooded mass at the transition between two major floristic regions, the Mediterranean and the Eurosiberian, both represented here (Polo i Alberti, 1987). But much of the diversity is due to successional changes, as the cork oak forest is in various stages of recuperation after having been exploited for the cork. Since cork is taken at 11-year intervals, and since some properties have not had recent extraction at all, there is a patchwork pattern of infinite complexity and varying rapidly from one property to the next. Because of this situation, comparable to that of the areas protected by Environmentally Sensitive Area (ESA) status in other parts of Europe, but not in fact covered by the ESA legislation, a management solution must be found through a combination of different interests, including wildlife conservation, countryside conservation and local economic regeneration.

A movement for conservation began with the publication in 1975 of a Manifesto of the Gavarres, a document subscribed to by a thousand people. Values evoked by the Manifesto were those of conservation of the environment against both tourism-related conversions of land use and conversion of the natural forest into eucalyptus plantations; protection of a microregion of historic interest with its own economy and culture; and protection of the old farm buildings and archaeological sites. The holding of a festive meeting in May 1976, in the heart of the Gavarres at Romanyà, a prehistoric sacred site, endowed the movement with spiritual values. Four thousand people supported what had become a substantial movement for a Nature Park at this time (Ventura, 1987). This broad swathe of interests contrasts with the previous phases and the simple protection of what has been termed above, for Phase III, wild nature. Part of the movement was against the effects of tourism, but growth of the domestic Spanish economy also elicited further reactions; an example is the active movement opposing the new bypass for Girona – the N 11 – which was proposed in 1983 and built by 1992. This clips the western margins of the Gavarres in skirting Girona city (Gispert, 1987). The Manifesto called for park status for the Gavarres and several other hill blocks of the region, including Montgrí and the Cap de Creus peninsula. This has yet to be achieved, despite continuous pressure. But the fight for the Gavarres has led to formation of a group Friends of the Gavarres – *Amics de les Gavarres* – which has continued the fight by regular meetings and lobbying of local politicians, and promoted in 1990 an extensive study of the Gavarres to establish the bases for its declaration as a National Park. Its members include many city people from Girona, as well as academics and scientists and people living in or around the hill mass. The very fact that this region can be even contemplated as a protected area indicates the changes achieved in attitudes towards rural land areas.

Some of the activities in and around the Gavarres which have been opposed by the Friends are not the classic forms of tourism – hotels and related services. In the municipality of Calonge as an example, the concerns include waste tipping by the municipality within the forest (now stopped officially, but still continuing clandestinely), proposals for residential construction and existing second-home-type development and the use of a piece of wasteland in the same municipality for cross-country motor-bike racing. All these activities are of course part of the more extensive tourism which has emerged post-1970. Nor has the group been entirely successful in checking the incursions; the recent development of a large golf-course on the coastal fringe of the forest mass is one encroachment which could not be stopped.

Not all reaction has been in opposition to tourism. In the proposals for a Regional Park from the Friends of the Gavarres (AAGA, 1992) there are recommendations for restoring the *masias*, the traditional farmhouses in the hill mass, so as to accommodate visitors and to develop rural activity holidays. The traditional cork manufacturing industry is also included in the study, and proposals are made for its revitalization and diversification away from basic bottle cork manufacture, so as to set up workshops where craftsmen might be seen at work with the raw material and items made from cork, including end-products which might be sold to visitors. Regeneration of cork extraction in the forest in combination with other farm activities is another recommendation. Cork extraction from the trees is uneconomic on its own, but it might play a role within a diversified farm economy. Clearly the intention is to continue to provide for tourists, but not for mass tourism of the coastal type. Instead, this would be a countryside tourism, a kind of rural heritage industry, which would combine with traditional land management of the Gavarres.

Looking briefly outside the Gavarres, the whole third phase of reaction to tourism is complex and uneven in its effects. Sant Feliu de Guixols provides an example, where the visible reaction has been to continue to support tourism. In this little town, as in many others, the population is of very varied origins because of the large number of second-home owners, many of whom come from other parts of Spain and even from foreign countries. As a result, it is difficult to mount a campaign of any kind based on local support. At Sant Feliu de Guixols (Morris, 1985), a pleasant little town central to the Costa Brava and long characterized by its cork processing industry, a new *Plan General de Ordenación Urbana* had to be produced but, as in most municipalities along the coast, there was no planning office so that the plan was formulated by Barcelona planners contracted about 1980. At the Catalonia region level, this work was regarded as of such high quality that it was published among a series from the Catalan Government Department of Urbanism (Pie i Ninot *et al.*, 1983). It was restrictive of tourism, calling for limits on hotel heights, improvement in the existing *urbanizaciones* so as to provide green spaces, building of museums of the old cork industry, but removal of the existing factories from the town. Local reception for the plan was less enthusiastic. When advanced to the stage of public exhibition for comment, it was fiercely contested, because it seemed to threaten tourism promoters in the town, and a few leading businessmen were able to orchestrate a strong opposition to it, on a variety of questionable arguments. One of the leading opponents was an estate agent who was able, through his firm, to mount petitions against the Plan from second-home owners who lived outside the area, often outside Spain, and had limited knowledge of the issues, but who were written to at their homes in Germany and the Netherlands. In this way, an opposition of very limited representativity was able to quash the Plan. It was approved later after modifications and submission to a controlling committee in Girona. However, this Plan has never been put into practice by the town council, fearful of the opposition which had been raised at the time of public inquiry. This case indicates the ambivalence of the local reaction to tourism; on the one hand, many people have opposed the permanent extension of tourism; on the other, where tourism business interests have been active, they have been able to manipulate public interest so as to maintain a developmentalist approach to tourism.

Some Final Points: Understanding the Changes

Over a period of 30 or 40 years, the Costa Brava has gone through a tourist development cycle and is attempting to revive the industry on a new basis. A contention of this chapter has been that alongside the production aspects of this development there has also occurred an evolution of local attitudes towards tourism, which attitudes are achieving equal importance with the intentions of the developers. It may be thought remarkable that attitudes could change so rapidly over one generation, and that some special explanations are needed beyond the expectable intergenerational shift in thinking throughout modern time.

One set of such explanations is provided by Recio (1992), who argues that, under Franco, a Spanish mode of production imposed a rapid growth of tourism within a generally developmentalist model in Spain. This political economy approach presents tourism as something which Catalans did not want, but which was imposed on them by the Spanish form of capitalism, highly centralized in Madrid, with Catalonia the dependent area. Once this dominance/dependence relation was loosened, even slightly, in the last years before Franco's death in 1975, ecological interests were allowed to surface. Their sudden appearance was thus effectively a reaction and a release of pent-up forces. From what has been written earlier, it is clear that such an argument is of very limited validity. Tourism has been shown, in this region at least, to have developed separately under private initiative with little help, let alone imposition, from central government.

In reality, the picture is more complex, as Spain has undergone more than one transition. Besides the departure of Franco, there has been the rapid urbanization and industrialization of Spain. From 1960, when well over 50% of the population lived in rural areas and 61.7% were directly employed in agriculture and other primary activities, the population has moved rapidly into the cities, with only 16.7% working the land by 1985. This itself has undoubtedly caused big changes in attitudes to the land. A peasant society and, perhaps even more, one which has in one generation been released from peasant status is unlikely to value the countryside highly. One which is already urbanized and industrialized can readily attribute romantic values to countryside which other people have to maintain. We can perhaps compare the rise of environmentalism and care for landscape at the present time in Spain to the rise in concern for the countryside in late nineteenth-century Britain, leading to such institutions as the National Trust in 1895. Alongside this great demographic, social and cultural transition in Spain, there has been the rapid rise of regionalism or local nationalism in different parts of Spain, the most intellectually advanced version of which is Catalanism, endowed with a rich cultural expression through the outpouring of literature in the Catalan language and formalized through regional government with substantial autonomous powers. Fighting against some of the consequences of 1960s tourism and fighting for the protection of local traditional features in the Catalan landscape have been a legitimate form of regionalism, ever since the late days of Franco, when this kind of protest was allowed as an activity which did not challenge Madrid and the central state politically. Territorialism is one of the fundamental forms of nationalism, and the argument here is that even the protection of small areas, when endowed with a rich cultural content, may be regarded as a derivative of such nationalism.

This is a different argument from that of Recio, who identified the economic aims of central government and counterposed them to local ones. What is proposed here is that the cultural and political sentiments are primary forces, alongside those value systems which relate to the urbanization and industrialization of the region, and that these forces asserted themselves after Franco's demise and became attached to the protection of nature, of landscape and of valued places with special historic meanings. An expression of the great importance of the traditional historical landscape and its structure of administration has been the debate over new administrative divisions

(Casassas and Clusa, 1981; Morris, 1990) and the creation of the new counties – *comarcas* – in Catalonia in the 1980s. In effect, this has been a restoration of very ancient units, in the pattern inherited from medieval time rather than based on any economic concepts such as urban hinterlands. As in the case of Britain, where the old counties have seen a revival in spite of all arguments about separate urban versus rural areas or combining large cities with their hinterlands, Catalonia has re-emphasized traditional county structure. The period from the time of Napoleon to the death of Franco may be regarded as an interlude when provinces and municipalities were the two local levels which were permitted. On the other hand, the evolution of social attitudes to tourism is likely to be a feature not confined to Spain and its special political circumstances. Because of greater levels of education and understanding, it probably warrants more attention by students of tourism development. The case made in this chapter is thus for a strong change process in the attitudes of locals towards the key tourism industry, from receptive to critical and selective. This does not precisely fit any tourist cycle, but follows a sequence of its own, coloured by the type of local culture and the changes which may occur in the political structure of the country. For sustainable development of a tourist region, it may be as important to know this kind of local reaction and its progress as that of the progress of industry itself.

References

AAGA (Associació d'Amics de les Gavarres i d'Ardenyà) (1992) *Les Gavarres: Anàlisi i Propostes de Gestió.* Diputació de Girona, Girona, mimeo, 218 pp.

Butler, R.W. (1975) Tourism as an agent of social change. In: *Tourism as a Factor in National and Regional Development.* Occasional Paper 4, Department of Geography, Trent University, Peterborough, Ontario, pp. 85–90.

Cals, J. (1974) *Turismo y Política Turística en España: una Aproximación.* Colección Laureano Figuerola, Ariel, Barcelona, 285 pp.

Cals, J. (1982) *La Costa Brava i el Turisme. Estudis sobre la Política Turística, el Territori i l'Hoteleria.* Kapel, Barcelona, 271 pp.

Casassas, L. and Clusa, J. (1981) *L'Organització Territorial de Catalunya.* Fundació Jaume Bofill, Barcelona, 326 pp.

COCIG (Cambra Oficial de Comerç i Industria de Girona) (ed.) (1978) *Debat Costa Brava: Ponències, Comunicacions i Documents de les Jornades Celebrades del 20 de novembre al 18 de desembre del 1976.* COCIG, Girona, 310 pp.

Doxey, G.V. (1975) A causation theory of visitor–resident irritants, methodology, and research inferences. In: TRA (Travel Research Association) (ed.) *The Impact of Tourism, Sixth Annual Proceedings of the Travel Research Association.* TRA, San Diego, pp. 195–198.

Esteban Noguera, J. (1978) Acció urbanitzadora a la Costa Brava. In: COCIG (Cambra Oficial de Comerç i Industria de Girona) (ed.) *Debat Costa Brava: Ponències, Com-*

unicacions i Documents de les Jornades Celebrades del 20 de novembre al 18 de desembre del 1976. COCIG, Girona, pp. 17–26.

Generalitat de Catalunya (1992) *Pla d'Espais d'Interès Natural*, 5 vols. Generalitat de Catalunya, Departament de Política Territorial i Obres Públiques, Departament d'Agricultura, Ramaderia i Pesca, Barcelona, 32 pp.

Gispert, E. (1987) La variant de la N II per l'est de Girona. In: *Les Gavarres: un Parc Natural Irrenunciable. Revista de Girona* 122, 69–70.

Jiménez Luna, P.A. (1983) *Legislación del Suelo*. Tecnos, Madrid, 581 pp.

Miossec, J.M. (1976) *Eléments pour une théorie de l'éspace touristique*. Les Cahiers du Tourisme, C-36, CHET, Aix-en-Provence, 63 pp.

Morris, A.S. (1985) Tourism and town planning in Catalonia. *Planning Outlook* 28(2), 77–82.

Morris, A.S. (1990) Local government in Catalonia: some problems of the new era. *Geoforum* 21(4), 411–419.

Morris, A.S. (1992) A sea change in Spanish conservation, with illustrations from Gerona province. *ACIS (Association for Contemporary Iberian Studies)* 5(2), 23–30.

Morris, A.S., Dickinson, G. and Priestley, G.K. (1989) *Land Use Change and Conflict in the Bay of Roses, Costa Brava*. Occasional Paper Series 25, Department of Geography, University of Glasgow, 22 pp.

Murphy, P.E. (1985) *Tourism: a Community Approach*. Methuen, London, 250 pp.

Ortega Hernández-Agero, C. (1989) *El Libro Rojo de los Bosques Españoles*. Adena–WWF España, Madrid, 389 pp.

Pearce, D. (1995) *Tourist Development*, 2nd edn. Longman Scientific and Technical, Harlow, 341 pp.

Pie i Ninot, R., Barba i Casanovas, R., Miró i Miró, R., Relea i Gines, F., Raurich i Puigdevall, J.M., Sicart i Orti, F. and Vilanova i Claret, J.M. (1983) *Castell-Platja d'Aro, Sant Feliu de Guixols, Santa Cristina d'Aro 1981*. Direcció General d'Urbanisme, Generalitat de Catalunya, Barcelona, 212 pp.

Planas i Roig, M. (1987) *Castelló d'Empúries*. Quaderns de la Revista de Girona No. 11, Diputació de Girona, Girona, 95 pp.

Polo i Alberti, L. (1987) La flora i la vegetació. In: *Les Gavarres: un Parc Natural Irrenunciable. Revista de Girona* 122, 24–27.

Recio, A. (1992) The ecologist movement in Spain – boomerang of the development model. *International Journal of Sociology and Social Policy* 12(4–7), 161–177.

Saurí Pujol, D., Ribas Palom, A., Breton, F. and Llurdés Coit, J.C. (1996) Estrategias tradicionales de aprovechamiento de espacios inundables, el sistema ampurdanes de les 'closes'. In: *Actas del XIV Congreso Nacional de Geografía, Salamanca, del 5 al 8 de diciembre 1995*. Asociación de Geógrafos Españoles, Madrid.

Tamames, R. (1977) *Introducción a la Economia Española*, 11th edn. Alianza, Madrid, 571 pp.

Urry, J. (1990) *The Tourist Gaze: Leisure and Travel in Contemporary Societies*. Sage, London, 176 pp.

Ventura, J. (1987) Historia d'una reivindicació collectiva. In: *Les Gavarres: un Parc Natural Irrenunciable. Revista de Girona* 122, 69–71.

6 Waterfronts, Tourism and Economic Sustainability: *the United Kingdom Experience*

J. ARWEL EDWARDS

Introduction

Water-related tourism has provided one of the principal growth areas of development and change over the past century. Until the 1960s, except for the very wealthy, this was essentially local in character but, since then, chartered jet aircraft and package holidays have opened up worldwide overseas destinations which offer the attractions of sun, sand and warm sea waters. Such tourism has two major impacts, one land-based, the other water-based. The most visible are the land-based impacts, especially at the waterfront interface: piers, promenades, hotels, apartments, camp grounds, bars, cafés, restaurants, while, at a further distance, are found parks, funfairs, golf-courses and theme parks. There is equal variety in the range of water-based activities: angling, jet-skiing, water-skiing and paragliding, snorkelling and subaqua, surfing and swimming, rowing, canoeing and yachting. In the United Kingdom, it was calculated that in the late 1980s such activities represented a substantial market of some 3 million people, who spent £520 million annually. During the past three decades or so this more general field of coastal and water-based tourism and recreational activities has experienced exceptionally strong growth in the specific area of yachting. Indeed, 'yachting has proved to be one of the boom sports of the post-war period' (Sports Council, 1991, p. 196) and associated with this, in the United Kingdom and elsewhere, has been the development and redevelopment of waterfronts, usually with a marina as a central focus. This chapter is concerned with examining the relationships between waterfront developments, tourism and economic sustainability. It begins by identifying a number of general issues and then proceeds to two contrasting case-studies, the one – Swansea – involving the redevelopment of abandoned docklands and organized through public-sector involvement, the other – Port Solent – being a *de novo* private-sector initiative. In both, the relevance of issues of economic sustainability are examined.

In one of the most comprehensive surveys of local authorities in the United Kingdom, Jones (1993) established that, in 1989, there were 221 waterfront redevelopment schemes in the process of completion. Of these, 181 were in England (extending over 8500 ha), 25 in Wales (1720 ha) and 15 in Scotland (368 ha). In aggregate, 41% of these projects had a tourism/ leisure function but, in Wales, this proportion rose to 88%. While one facet of this tourism is related to and dependent upon visiting yachts, tourism has provided a much greater stimulus through attracting land-based visitors. Here, so-called 'third-phase' multifunctional developments, which incorporate 'festival-market', residential, retail, commercial and other tourism dimensions, have assumed increasing prominence, as described by Mann (1988):

> One of the keys to its success (i.e. the multi-functional complex) is the way in which dining, entertaining, boating and sailing, artists' and artisans' studios and workshops, and all the other central and incidental uses that make up a 'people-place' were made to work at the heart of the property even before much else of the residential and office development came along. (p. 180)

Examples include the Baltimore, USA, waterfront, Port Grimaud, France, and Port Solent, England.

Gates (1987) has suggested that the general trend to exploit and revive the nation's abandoned dock areas will increase as coastal leisure provision becomes saturated, while Martin and Jones (1988) have noted the importance of tourism and recreation in the waterfront paradigm. In this context, it is increasingly recognized that the term 'environment' should be extended more widely to encompass 'people, their creations and the social, economic and cultural conditions that affect their lives' (Duffield and Walker, 1984). This clearly has application to the constructional impacts of redeveloped old docklands, *de novo* waterfront developments and marina schemes which may or may not be incorporated within the above. That waterfront developments have major environmental impacts has been recognized by Hampshire County Council in south-central England – the county in which the second case-study is located – which now opposes the construction of additional marinas, since it considers that its coastlines cannot absorb further developments. For example, in 1976 the Solent area of Hampshire had 25,934 boat and 17,018 wet moorings but by 1993 these had increased to 32,970 and 24,520 respectively (HCC, 1993a). Another more specific environmental aspect is found in the attention being given to general pollution and water quality issues caused by large numbers of yachts in and around marinas (Nece and Layton, 1989). More recently, the term environment has been linked to the issue of sustainability, defined in the Brundtland Report (WCED, 1987) as 'development that meets the needs of the present without compromising the ability of future generations to meet their own needs', while Gibbs (1994) has linked the themes of environment, sustainability and urban economic development – which are the context for most waterfronts – in the following ways:

> Environmentally sustainable urban economic development can be defined as local economic change which contributes to global environmental sustainability, while also enhancing the local natural and constructed urban environment. (pp. 106–107)

> Sustainable development favours increased local control over development decisions, and such 'bottom up' development strategies would require devolution of more decision-making authority to the local level, particularly in the UK. (pp. 106–107)

> The concept of sustainable development is increasingly being discussed as forming the basis for economic development. However, there has been little discussion of the consequences of this for either national or local economies. (p. 99)

The relationships between tourism and local economic sustainability have been clearly identified by Owen (1991), who, drawing upon the experience of working for more than two decades for the Wales Tourist Board, has produced a check-list of key principles which underlie sustainable tourism.

1. It recognizes that tourism is a potent economic activity which brings tangible benefits to the host community as well as to the visitor; however, it accepts that tourism is not a panacea and must form part of a balanced economy.
2. Its scale and the pace of development respect the character of the area; it is value-conscious and seeks to offer a high-quality experience; customer care is important.
3. It seeks to bring optimum long-term community benefit in the form of varied, attractive and well-paid jobs; it is less concerned with short-term speculative gain for only a few.
4. It is sensitive to the needs and aspirations of the host population; it provides for local participation in decision-making and for the employment of local people.

The final theme of this chapter and its case-studies is that of the guidelines of planning and sustainability within which marina developments must take place. In the United Kingdom the general framework is set down by central government in its Planning Policy Guideline document PPG1 (DoE, 1991), which states:

> The Government has made clear its intentions to work towards ensuring that development and growth are sustainable. It will continue to develop policies

consistent with the concept of sustainable development. The planning system, and the preparation of development plans in particular, can contribute to the objectives of ensuring that development and growth are sustainable. The sum total of decisions in the planning field, as elsewhere, should not deny future generations the best of today's environment.

This study examines two case-studies of the marina-focused waterfront development in the counties of Hampshire and West Glamorgan. Hampshire County Council has produced a specific document on *Sustainability* as part of its Structure Plan Review (HCC, 1993b), while its coast strategy document (HCC, 1991) recognizes the fragility of its coastal lands. There is clear evidence that Hampshire County Council argues that the coastline – and local residents – cannot sustain further coastal marina development since its documents state (HCC, 1991, p. 51) that the Council 'considers that the Solent's ability to absorb further recreational activity safely and without damage to the environment has been stretched almost to the limit. The provision of new facilities should be treated with great caution.' Thus the Council will oppose the provision of new marinas, beyond existing commitments. In accordance with this, the City has adopted Policy C10, which states that, on the coast:

- permission for development will not normally be granted if the site is an undeveloped coast, or is in an area of nature conservation value, including intertidal areas;
- if the site is on a developed part of the coast, including existing boatyards and marinas, permission will normally only be granted for development for uses requiring access to the shore.

West Glamorgan County Council has only one marina within its coastal zone and so experiences little pressure for development in comparison with Hampshire. Its Strategic Structure Plan also incorporates the requirements of sustainability laid down by PPG1, noted above, including:

- Policy EQ1:
  Encouragement will be given to development proposals in appropriate locations that:
  (i) reduce the need for the consumption of energy and non-renewable resources, or
  (ii) make effective use of renewable resources or waste products, or
  (iii) reduce or avoid pollution and other dangers to life, or
  (iv) demonstrate significant environmental benefits, including the restoration and aftercare of degraded sites to a satisfactory standard.
- Policy C11:
  Opportunities for regenerating areas of derelict dockland and despoilt coastlines for new uses and activities will be encouraged where appropriate (WGCC, 1993).

The Waterfront Case-studies

Case-study 1: Swansea Maritime Quarter

Swansea Maritime Quarter was initiated in the mid-1970s by the City Council. It can be seen not only as an example of the second type of marina development outlined in the environmental policies above, i.e. marina-focused waterfront redevelopment, but also of the 'bottom-up' economic development strategy of sustainable development identified by Gibbs (1994). In this respect, it contrasts strongly with the 'top-down' approach of the regeneration of such United Kingdom docklands as London, Merseyside, Teesside, Tyne and Wear and Cardiff Bay through the Urban Development Corporation initiatives of central government.

In 1974 Swansea Council turned its attention to the possibilities of regenerating the old and abandoned South Dock (Fig. 6.1), originally opened in 1859 but closed to shipping in 1969. Proposals were then made for it to be filled in for either industrial estates or an inner urban relief road but this did not happen although it was partly infilled. In 1974 there were 1894 people working in 83 firms in the Maritime Quarter (Table 6.1) and most employment was full-time and male-orientated (Edwards, 1987).

In 1974 the Council:

> looked at this derelict part of Swansea's inner city as an exciting challenge, seeing derelict dock basins, redundant buildings and old quay sides not as a problem but rather as a combination of opportunities. With vision and in combination with new developments they could add brand new economic, social and environmental dimensions to the city.
>
> (Swansea, 1975).

One of the Council's first actions was to acquire the ownership and leases of most buildings and land around the Dock so that comprehensive and integrated development planning could take place. This enabled the Council to embark on a wide range of land- and water-based improvement

Table 6.1. Firms and employment in Swansea Maritime Quarter, 1974–1986 (from Department of Employment, 1974, 1976, 1981, 1986).

| | DEVELOPMENT AREA | | | | | | CONSERVATION AREA | | | | | |
| | No. of firms | Employment | | | | | No. of firms | Employment | | | | |
Year		MFT	MPT	FFT	FPT	Total		MFT	MPT	FFT	FPT	Total
1974	35	807	17	200	77	1101	48	602	45	175	61	883
1976	17	402	15	142	42	601	49	575	41	147	63	722
1981	6	216	0	77	26	319	53	624	41	144	101	665
1986	28	289	44	153	52	538	66	432	103	187	176	619

MFT, male full-time; MPT, male part-time; FFT, female full-time; FPT, female part-time.

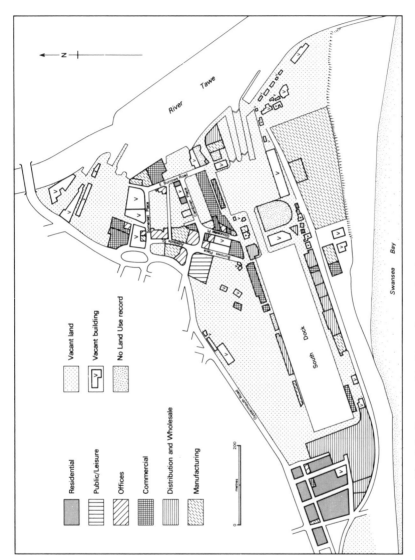

Fig. 6.1. Swansea Maritime Quarter land use, 1974 (from Edwards, 1987).

schemes in order to improve the general natural and built environment of the area as part of a wider programme of inner urban renewal. Secondly, the Council used its planning powers to link the Maritime Quarter with the retail and commercial centre of the City through direct physical routes – pedestrian walkways – and by providing such attractions as museums, leisure/recreation facilities to attract people into the Maritime Quarter. The Council was seeking to improve the local economy in the face of rising unemployment and a declining population. Between 1980 and 1986 a total of 36,492 workers were made redundant within the surrounding county of West Glamorgan and local unemployment rose from 8% in January 1980 to 16% in December 1981. Further, both County and City were losing population: that of the county fell by 1.5% between 1971 and 1981 and by 2.9% between 1981 and 1991, while corresponding declines in the City were 3.3% and 1.0%. The local economy was deindustrializing rapidly and any new employment generated by the marina would help ameliorate the local economic crisis. So, in common with numerous similar developments elsewhere, the 'marina waterfront village' paradigm appeared to offer a means of dealing with numerous pressing economic as well as natural and built environment concerns.

The initial impacts of the Council were both negative and positive. In an economic context, the initial impacts were negative in that the process of local land clearance around the Dock resulted in the disappearance of 29 businesses by 1981, of which ten ceased trading completely and 19 were re-established elsewhere (Edwards, 1988). On the other hand, temporary employment rose dramatically during the 1980s as rebuilding began and, by 1987, the building industry had added the equivalent of 931 permanent jobs to the local labour market (Williams, 1995). By 1986, the Maritime Quarter was generating new jobs, though not equal in numbers to those of 1974, and many of these were part-time and female. Further positive impacts could be seen in the clearance of materials from the water area of the Dock, the improvement in water quality, the removal of numerous old buildings and the beginnings of construction of new residential apartments, retail outlets, cafés, bars and restaurants.

A second aspect of sustainability, noted earlier, relates to the 'local control over development decisions ... [and] ... devolution of more decision-making authority to the local level' (Gibbs, 1994, pp. 106–107). The commitment of Swansea City Council was expressed in the *Interim Planning Statement* (Swansea, 1975). Amongst its policies it argued that local authority provision should be made for the conservation and improvement of old buildings of character, attracting and maintaining the interest of tourists and visitors to the area and enhancing the visual quality and interest of the redeveloping Maritime Quarter. In his assessment of the role of Swansea City Council, Williams (1996) writes:

In view of the protracted financial timescale it can be argued that only a public agency such as a local authority could be expected to meet the challenge presented by the redundant South Dock. Such a view represents a departure however from the contemporary view of Central Government as manifested in the introduction of the Urban Regeneration Grant process in 1987, this grant being designed specially to by-pass the function of local authorities, by giving direct support and encouragement to private companies planning to undertake redevelopment projects of major proportions. The response is likely to materialize in economically buoyant regions . . . [but] . . . in regions of economic decline, however, where the process of urban change is seen as making a valuable contribution to the restructuring of the economic base, the role of the public agency is critical. Swansea can justifiably claim to be at the vanguard of such a system, and in the work undertaken in relation to the Maritime Quarter can be seen as foreshadowing the highly acclaimed process of setting up Development Corporations to influence and direct urban change.

Between 1974 and 1990 Swansea City Council spent £24.2 million in acquiring, clearing, providing infrastructure for, conserving and enhancing the Maritime Quarter. This, in turn, generated £51.3 million of expenditure from elsewhere – private funds, grants from agencies – a ratio of 1:2.2. By 1990, these investments were producing an annual property rental to the City of £0.45 million and taxes of £0.99 million. In addition, the boat owners moored in the marina basins bring in considerable income to the local economy (Edwards, 1985). His survey showed that boat owners alone benefited the local economy by £0.86 million in that year from berthing charges and other expenditures – food, drink, entertainment, chandlery purchases, servicing fees.

A further aspect of the development of the Maritime Quarter is its focal role in the provision of leisure facilities and the promotion of tourism for the City. This is done in a variety of ways – by developing a leisure centre, museums, commercial facilities, cafés, bars, restaurants, shops and art galleries, permanent and visiting historic ships, and the marina, with support facilities. In the early 1990s the marina concept was extended by the construction of a barrage across the mouth of the River Tawe. All the above are surrounded by high-quality buildings, street landscapes and a large number of commissioned sculptures. The success of these developments emerged from a 1992 visitor survey (Jones, 1993). Just over 40% came from the City of Swansea and a further 11% from the surrounding county – a not dissimilar profile to other tourism studies. Almost one-half of all visits (49%) were for informal leisure facilities – sightseeing, walking, shopping, casual eating and drinking – but boating/marina-related activities accounted for only 3.5%. At least 32% of visitors stated that they went to the Maritime Quarter at least once a week, which indicates that the waterfront and its activities provide a well-established attraction. Many waterfront developments have attracted unfavourable criticism because of their poor design, lack of character and generally unimpressive environments. In Swansea,

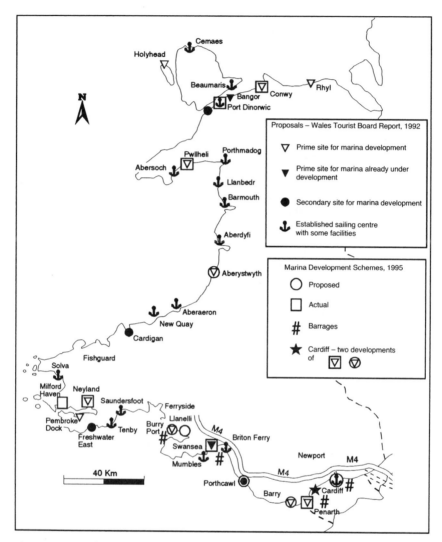

Fig. 6.2. Coastal marina proposals (1982) and developments (1995) in Wales.

however, surveys show that visitors have a high regard for the physical appearance and environmental quality and there were favourable responses to the marina harbour, housing, cafés/pubs and sea-front promenade.

Finally, the success of the above marina has had wide-ranging repercussions throughout the country. In Wales, where there were no marinas prior to that in Swansea, the above initiative prompted the Wales Tourist Board (WTB, 1982) to commission a study of potential for marina development. This has provided the blueprint for planning since and, by 1995, five additional marinas had been developed (Fig. 6.2).

Case-study 2: Port Solent waterfront multifunctional complex

This differs from Swansea in that it is a totally new development and is not part of a regeneration project of an old waterfront. It is, however, similar in that it involves the recycling of land from old to new uses, but in this case as part of an old refuse site of 193 ha, which has been reclaimed and is ventilated by a sophisticated system of methane gas extraction flues. It contrasts with Swansea in that it is a privately financed development involving a consortium of companies. Its location is also different, for it is situated in the most intensively developed coastal region of the United Kingdom for yachting and marina development. In order for it to succeed, it is essential that it surpasses other south coast waterfront and marina developments in design, layout, facilities and environmental attractiveness. Construction of the £200 million privately financed Port Solent multifunctional complex in the northeastern corner of Portsmouth Harbour began in January 1986. Planned as an integrated whole, it contains marina, residential, commercial, retail, office, entertainment and boat-servicing/repair components.

The marina opened for use in April 1988 with 304 berths occupied, on target for an eventual 850-berth locked basin divided between 470 non-residential and 380 residential berths (Fig. 6.3). The residential sector will eventually have 414 one-, two- and three-bedroom houses and 168 apartments but the economic recession has slowed down the pace of construction. In 1989 the commercial sector began to come 'on stream'. The Boardwalk is an extremely attractive retail/entertainment complex, of innovative design

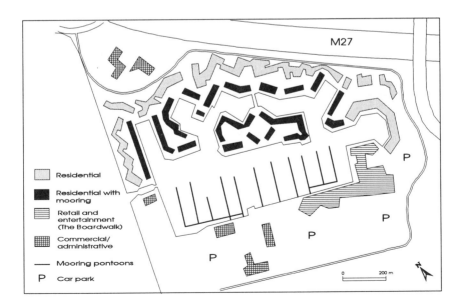

Fig. 6.3. Port Solent waterfront development, 1995.

and layout. It has 22 retail units, five restaurants, a wine bar, a large pub, nine office suites and seven apartments. An additional commercial phase is provided by the Slipway complex of 36,000 square feet of commercial space for boat sales, chandlery and leisure-related businesses, while a large office development was completed at the landward entrance to Port Solent and occupied. In 1989, the Waterside Club provided an additional social, eating and entertainment facility and its popularity was such that it quickly exceeded 1500 members, many of whom were not interested in sailing but were attracted by its qualities of service, food and beverages. As part of this initial phase of development, a Port House was constructed to administer the marina and to provide office space for a flotilla holiday company. By 1992, it was evident that this waterfront development was attracting visitors from a wide area (see below), such that, in December 1992, an additional entertainment facility was provided by a new ten-screen multiplex cinema next to the Boardwalk, a major boost for visitor numbers.

Surveys of the entire Port Solent waterfront complex have been under-taken since 1990 to determine the success of its operations and its appeal to tourists and visitors. First, the catchment areas of both water- and land-based visitors have increased substantially since opening in 1988. Boat owners are drawn from a wide area, including the west Midlands, Home Counties and London, and, in 1993, around one-quarter of these were visit-ing boat owners. Many of these have high disposable incomes and are increasingly demanding of high-quality services and facilities – and here questionnaire surveys showed that Port Solent ranked highly in comparison with other marinas along the south coast, such as Brighton, in the UK, and elsewhere in Europe, for example, St Malo and Cherbourg in France.

The number of land-based visitors had grown to more than 200,000 per annum by 1991 – as measured by footfall trip-monitors – and the addition of the new multiplex cinema has increased this. The Boardwalk shopping, eating/drinking and entertainment area provided the principal focus of attrac-tion within the wider waterfront complex and its success can be gauged from the fact that the proportion of visitors coming from more than 26 miles almost doubled between surveys undertaken in 1991 and 1993. Clearly, the objective of further developing the economic viability of the complex, as measured by these criteria, is proving effective. This is reflected in the high levels of visitor satisfaction. A 1993 survey showed that almost one-third visited the complex weekly and over 20% at least once a fortnight. Almost one-third of visitors stated that they would spend more time at the complex over the next 6 months and this indicates its ability to enhance its power of attraction. Reasons cited for visiting were fourfold: almost equal proportions – 36% – came to shop, eat and drink; 14% came to browse over the boats, the shops and the overall development; and 11% attended the cinema.

This multifunctional and integrated waterfront development was described by numerous tourists as 'superb', 'exciting' and 'innovative', and the single most important message to come from the survey was that it was regarded as excellent in concept and had few peers in its market sector. In severe competition with numerous other waterfront and marina developments, it had created qualities of natural and built environment, hard and soft landscaping, that should sustain its future growth and development.

Conclusion

The issues discussed in this chapter, while United Kingdom-focused, are of general European concern, since waterfront development is now taking place across the entire continent. Swansea illustrated the refurbishing of older port facilities, now occurring from Amsterdam to Barcelona, while Port Solent exemplified the many *de novo* developments. Irrespective of origin, ownership or location, the influences of tourism, leisure and recreation are prominent. Since the early 1960s, water-related activities have become the greatest generator of tourism and recreation and have provided a spur for economic development and growth. This is true not only for new destinations which were established on formerly undeveloped sites, but also for built-up waterfronts where traditional industrial activities were in decline. The phenomenon of such postindustrial redevelopment has increased principally since 1980. In Swansea the redevelopment of an old dock has enabled new and progressive development approaches to be adopted. Here, the local authority has been the principal agent of development, using legislation to clear polluted waters, despoiled landscapes and old buildings. Innovative housing layouts, new industrial and commercial developments, the commissioning of numerous works of outdoor sculpture, new museums and other attractions for visitors have generated a new and exciting quarter in the city and assisted a region experiencing considerable economic distress. In particular, the municipal authority has used practices of local control over development to ensure a 'bottom-up' approach.

The Port Solent case-study provides a contrasting developmental framework of private-sector funding and control within the framework of local authority planning guidelines. Its location, in part upon the site of a former household refuse dump, provides an excellent example of land recycling which has helped to enhance the natural environment, while the quality of its built environment has attracted much favourable comment. For many, it exemplifies the late twentieth-century trend of breaking down the separation between home and work and between shopping and leisure and recreation.

Both case-studies are illustrative of the three key principles of sustainability discussed earlier, namely, 'of environmentally sustainable urban economic development ... enhancing the local natural and constructed urban environment' (Gibbs, 1994, pp. 106–107), of 'sustainable development favouring local control over development [through] bottom up development' (Gibbs, 1994, pp. 106-107) and, finally, of the main principles of economic development underlying tourism as identified by Owen (1991).

References

Department of Employment (1974, 1976, 1981, 1986) *Census of Employment*. HMSO, London.

DoE (Department of the Environment) (1991) *PPG1 – General Policy and Principles*. HMSO, London, 36 pp.

Duffield, B.S. and Walker, S.E. (1984) The assessment of tourism impacts. In: Clark, B.D. and Gilad, A. (eds) *Perspectives on Environmental Impact Assessment*. Reidel, Dordrecht, pp. 479–516.

Edwards, J.A. (1985) *The Identification and Location of Potential Users of the Swansea Yacht Haven in the West Midlands*. Report to the Economic and Social Research Council, Swindon. Department of Geography, University of Wales, Swansea, 40 pp.

Edwards, J.A. (1987) Marina quarter redevelopment: the case-study of Swansea. *Cambria* 14, 147–162.

Edwards, J.A. (1988) Public policy, physical restructuring and economic change: the Swansea experience. In: Hoyle, B.S., Pinder, D.A. and Husain, M.S. (eds) *Revitalising the Waterfront*. Belhaven Press, London, pp. 129–145.

Gates, J. (1987) Marinas – new wave in leisure. *Chartered Surveyor Weekly* 20(16 July), 24–25.

Gibbs, D. (1994) Towards the sustainable city. *Town Planning Review* 63(1), 99–109.

HCC (Hampshire County Council) (1991) *A Strategy for Hampshire's Coasts*. HCC, Winchester, 96 pp.

HCC (1993a) *Hampshire County Structure Plan (Review)*. HCC, Winchester, 118 pp.

HCC (1993b) *Sustainability*. HCC, Winchester, 12 pp.

Jones, A.L. (1993) Contemporary issues in waterfront regeneration – a case study of the Swansea waterfront. MPhil thesis, University of Wales, Swansea, United Kingdom.

Mann, R. (1988) Ten trends in the continuing renaissance of waterfronts. *Landscape and Planning* 16, 177–199.

Martin, T. and Jones, C. (1988) Urban waterfront development in the United States. In: *Watersite 2000*. Bristol City Council, Bristol, pp. 19–30.

Nece, R.E. and Layton, J.A. (1989) Mitigating marina environmental impacts through hydraulic design. In: Blain, W.R. and Webber, N.B. (eds) *Marinas: Planning and Feasibility*. Computational Mechanics Publications, Southampton, pp. 435–450.

Owen, R.E. (1991) Strategies for sustainable tourism: the theory and the practice. Unpublished paper presented to the *International Conference on Tourism: Development, Trends and Prospects in the 90s, 16–18 September, Kuala Lumpur*, 12 pp.

Sports Council (1991) *A Digest of Sports Statistics for the UK*. Sports Council, London, 240 pp.

Swansea (1975) *South Dock – Interim Planning Statement*. City of Swansea, Swansea, 6 pp.

WCED (World Commission on Environment and Development) (1987) *Our Common Future (Brundtland Report)*. Oxford University Press, Oxford, 400 pp.

WGCC (West Glamorgan County Council) (1993) *West Glamorgan Structure Plan, Review No. 2*. West Glamorgan, Swansea, 117 pp.

Williams, D. (1996) The South Dock, Swansea: a process of waterfront renewal. MPhil thesis, University of Wales, Swansea, United Kingdom.

WTB (Wales Tourist Board) (1982) *Harbour and Marina Study – Wales*. WTB, Cardiff, 117 pp.

7 Structural Dynamics of Tourism and Recreation-related Development: *the Catalan Coast*

GERDA K. PRIESTLEY

Introduction

It is a well-known fact that the provision of facilities associated with tourism development transforms the natural environment, and that this process can modify and even eradicate the original source of attraction. This is especially so in the case of sun/sand/sea tourism, originally attracted to unexploited and unspoilt beaches. As popularity increases, the hinterland of the beaches is built up, the beaches become overcrowded and the sea suffers the polluting effects of waste disposal and boat fuel. As accommodation provision increases, so too do conflicts. Obviously developers, in search of personal gain, favour expansion and municipal councils usually welcome the possibility of increasing revenues from sources such as building permits, property taxes and vehicle parking charges. However, the costs are high: environmental deterioration, overcrowding, traffic congestion, pollution and, eventually, even the loss of economic viability. It is perhaps understandable that resorts which developed during the first phase of expansion of mass tourism, in the 1960s and early 1970s, were allowed to grow almost indefinitely. It is not, however, so understandable that this trend should continue to the present, in the light of past experience.

Coastlines which are suitable for tourism and recreation development are not only a highly limited resource, but also an extremely fragile and vulnerable one. Moreover, in addition to the excessive pressure exerted directly upon them, they serve as the receiving zone for waste products from their inland hinterlands. In many cases, these areas suffer from the problems of marked seasonal variations in use, from almost total abandonment in the low season to gross overcrowding at the period of maximum demand. In fact, the coast has become the paradigm of the contradictions between development and environmental conservation. Obviously, the present pattern of development must be reconsidered in a search for a more equal balance between economic activities and natural resources.

The objective of this chapter is to identify the transformations which have taken place on the coast of Catalonia as a result of tourism and recreation development, to pinpoint the conflicts and problems that have consequently arisen and to indicate the solutions which have been or could be implemented in an attempt to reconcile development and the environment, in the light of this experience.

The Natural Environment

The coast of Catalonia extends for 595 km – 8% of the Spanish total – from the French border to the Ebro River delta. Morphologically, it is a very varied coastline (Figs 7.1 and 7.2). There are 302 km of beach – 15.2% of the Spanish total – which is slightly more than half the entire length of the coast. In the picturesque northern sector – Costa Brava – steep, rocky promontories alternate with relatively inaccessible coves and longer stretches of beach. The central section – Costa del Maresme – consists mainly of narrow beaches, which are highly vulnerable to erosion in winter storms, backed by a narrow coastal plain. The railway line and the main road border the coast along most of it, making access from Barcelona easy, but forming a barrier between the beach and its hinterland. To the south of Barcelona, the wide, sandy Llobregat delta beaches, backed by dunes and lagoons, give way to the cliffs of the Garraf Massif. Beyond, long, wide, sandy beaches with flat hinterlands, which were originally mostly swampy and therefore deserted, predominate on the remaining 150 km of coastline – Costa Daurada and Costa del Camp de Tarragona. Nevertheless, all of the beaches together occupy only 3285 ha, a mere 0.1% of the total surface of Catalonia.

Tourism and Recreation Development: Facts and Figures

Over a period of approximately 40 years, facilities for international mass tourism have developed in Catalonia, mainly concentrated on the coast (Fig. 7.3). In addition, the construction of second homes, mostly for Catalan city-dwellers, gained momentum after 1970 (Fig. 7.4). This phenomenon has greatly intensified the occupation of the coastal fringe, where the vast majority of such homes have been located. The importance of Spain as a tourism destination is well known, and Catalonia is one of the principal destination regions within the country. The main characteristics of Catalan tourism are outlined in Table 7.1, confirming not only the importance of

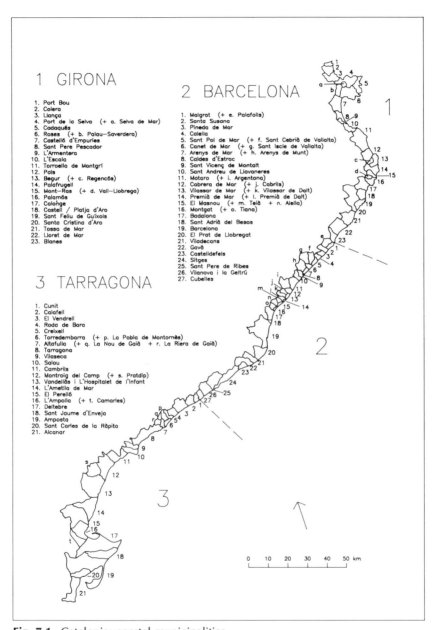

1 GIRONA

1. Port Bou
2. Colera
3. Llança
4. Port de la Selva (+ a. Selva de Mar)
5. Cadaqués
6. Roses (+ b. Palau—Saverdera)
7. Castelló d'Empuries
8. Sant Pere Pescador
9. L'Armentera
10. L'Escala
11. Torroella de Montgrí
12. Pals
13. Begur (+ c. Regencós)
14. Palafrugell
15. Mont—Ras (+ d. Vall—Llobrega)
16. Palamós
17. Calonge
18. Castell / Platja d'Aro
19. Sant Feliu de Guíxols
20. Santa Cristina d'Aro
21. Tossa de Mar
22. Lloret de Mar
23. Blanes

2 BARCELONA

1. Malgrat (+ e. Palafolls)
2. Santa Susana
3. Pineda de Mar
4. Calella
5. Sant Pol de Mar (+ f. Sant Cebrià de Vallalta)
6. Canet de Mar (+ g. Sant Iscle de Vallalta)
7. Arenys de Mar (+ h. Arenys de Munt)
8. Caldes d'Estrac
9. Sant Vicenç de Montalt
10. Sant Andreu de Llavaneres
11. Mataro (+ i. Argentona)
12. Cabrera de Mar (+ j. Cabrils)
13. Vilassar de Mar (+ k. Vilassar de Dalt)
14. Premià de Mar (+ l. Premià de Dalt)
15. El Masnou (+ m. Teià + n. Alella)
16. Montgat (+ o. Tiana)
17. Badalona
18. Sant Adrià del Besos
19. Barcelona
20. El Prat de Llobregat
21. Viladecans
22. Gavà
23. Castelldefels
24. Sitges
25. Sant Pere de Ribes
26. Vilanova i la Geltrú
27. Cubelles

3 TARRAGONA

1. Cunit
2. Calafell
3. El Vendrell
4. Roda de Bara
5. Creixell
6. Torredembarra (+ p. La Pobla de Montornès)
7. Altafulla (+ q. La Nou de Gaià + r. La Riera de Gaià)
8. Tarragona
9. Vilaseca
10. Salou
11. Cambrils
12. Montroig del Camp (+ s. Pratdip)
13. Vandellòs i L'Hospitalet de l'Infant
14. L'Ametlla de Mar
15. El Perelló
16. L'Ampolla (+ t. Camaries)
17. Deltebre
18. Sant Jaume d'Enveja
19. Amposta
20. Sant Carles de la Ràpita
21. Alcanar

0   10   20   30   40   50 km

Fig. 7.1. Catalonia: coastal municipalities.

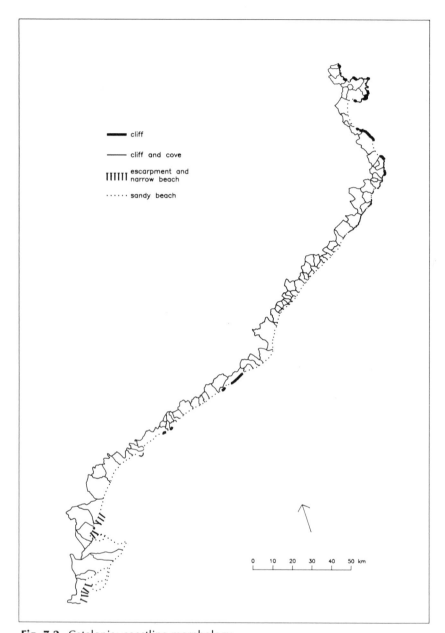

Fig. 7.2. Catalonia: coastline morphology.

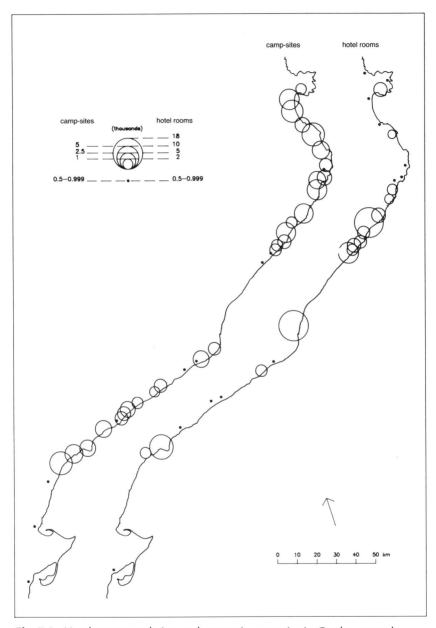

Fig. 7.3. Hotel accommodation and camp-site capacity in Catalan coastal municipalities in 1994 (based on Generalitat de Catalunya, 1994a, b).

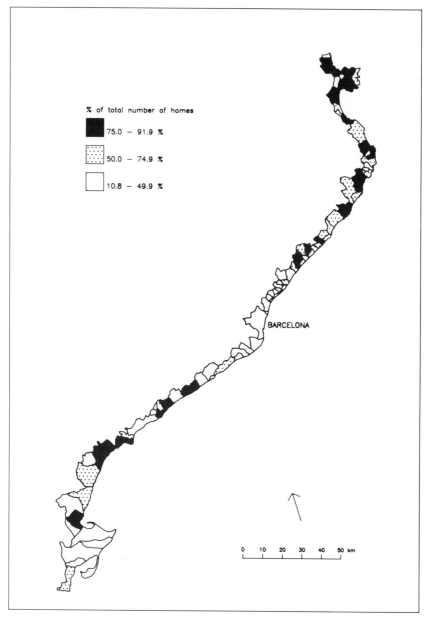

Fig. 7.4. Second homes in Catalan coastal municipalities in 1991 (based on INE, 1992).

Table 7.1. Tourism in Catalonia: some basic statistics (from INE, 1992; Generalitat de Catalunya, 1993; MCT, 1994).

	Total for Catalonia	% of Spanish total
Economic data 1992		
Gross income	$21,035 million	
Expenditure	$4,749 million	
Net income	$16,286 million	
Tourism accommodation		
Hotel beds 1992	229,568	5.1
Camp-site capacity (places) 1992	246,870	41.0
Private accommodation (beds) 1991	1,664,564	
Tourism demand 1992		
Total no. of tourists	14,822,406	28.6
Foreign tourism 1993		
Hotel visitors	2,483,220	19.2
Hotel bed-nights	11,799,111	14.2
Camp-site bed-nights	2,846,755	49.2
Domestic tourism 1993		
Hotel visitors	2,480,894	13.3
Hotel bed-nights	7,397,445	13.6
Camp-site bed-nights	6,700,625	54.1

Catalonia within the Spanish tourism structure, but also the importance of tourism in the Catalan economy.

The resulting impact on the environment was obviously inevitable. However, what these statistics do not reveal is the fact that the form and extent of impact vary greatly along the coast. There is a tendency to consider Catalonia, or even smaller sectors, such as the Costa Brava, as uniform units. Reality is very different: the Costa Brava includes well-preserved villages such as Cadaquès and major tourism destinations such as Lloret de Mar, which has 14.5% of all Catalonia's hotel beds. A more detailed analysis is therefore necessary in order to understand the true dimension of the impacts of tourism and recreation development (Miguelsanz, 1983). The diversity of the natural environment is obviously a conditioning factor, but it is also essential to trace the early evolution of tourism development in the region in order to understand the pattern of subsequent development and impacts.

The Pattern of Early Development

Development has followed different patterns and different rates in each municipality, as a result of the unequal operation of several recurring factors. Most early development was piecemeal, depending largely on small-scale investment on a local scale. For example, local inhabitants rented out rooms or even their entire apartment, added a floor on to their home and let it out in summer, expanded existing guest-houses, often using 'annexes', which, in some cases, were spare rooms in neighbours' houses. The original initiative seldom came from elsewhere; outside investors were generally attracted once a resort began to become established. It was at this stage that municipal policy began to play a part. However, rather than defining a long-term development policy, municipal councils tended to evaluate individual projects proposed, approving or rejecting them on a one-by-one basis. Policy therefore emerged slowly on an *ad hoc* basis as one decision followed another.

Paradoxically, the first two resorts to develop – Tossa de Mar and Sitges – are among those which have suffered the least transformations. In the case of Tossa, it was the filming of *Pandora* on location there in the early 1950s that aroused interest internationally in the Costa Brava in general and Tossa in particular. However, successive town councils resisted offers of large-scale tourist projects. As a result, Tossa has remained largely unaltered, whereas neighbouring Lloret de Mar has developed as one of the major destinations for mass tourism on the coast. At the time of the expansion of mass tourism, Sitges was already a renowned holiday resort, so further development required little stimulus. In fact, in 1960 it was calculated that of all foreign currency exchanged in Spain 1.8% was exchanged in Sitges. However, the rejection of an ambitious marina project in 1964, together with the constant refusal to allow high-rise development on the 1.5-km-long beach front to replace the detached houses built before 1940, constituted the effective refusal to develop as a mass tourism destination.

Natural Conditions as a Factor in Development

In the case of potential resorts where local initiative was lacking at the outset, natural conditions have been important in determining the extent and type of subsequent development. A sizeable beach and a considerable extension of surrounding land suitable for construction were necessary to encourage large-scale development. Such is the case of Platja d'Aro, now one of the largest resorts on the Costa Brava, but hitherto a deserted beach backed by a pine copse. Likewise, in Salou and Sant Antoni de Calonge, similar natural conditions permitted the widespread construction of accommodation for tourist purposes.

Accessibility was also an important factor. In many cases, difficulties of access coincided with the natural restrictions imposed by the small size of coves backed by steep pine-covered slopes, especially on the Costa Brava. For these reasons, several picturesque coves including Fornells and Sa Tuna –

in the municipality of Begur – and Tamariu – in the municipality of Palafru-gell – have remained relatively unspoilt. Proximity to an airport was an additional factor which encouraged the development of specific resorts. This is especially so in the case of Salou, just 10 kilometres from Reus airport, formerly used exclusively for military purposes. The development of Girona airport has also contributed to the expansion, but not the original develop-ment, of the major Costa Brava resorts.

Increased accessibility as a result of the improvement of the road net-work, and especially the construction of motorways, has had noticeable effects on the pattern of development. Easy access by road from France has encouraged the development of Roses, formerly isolated on the northern fringe of the Costa Brava. However, the importance of this factor is most clearly visible in the development of resorts on the Costa Daurada and the Costa del Camp de Tarragona. Poor accessibility prior to the construction of the coastal motorway during the 1970s had left the entire coast largely unspoilt, but as the motorway was extended, so too did the construction of second homes and camp-sites, gradually spreading southwards along the coast (Fig. 7.5). This was the era of the generalized acquisition of second homes by the rapidly expanding middle class in the major Spanish cities, and large extensions of land at low prices were available on this coast. The construction of hotels for the international tourist market was concentrated in already established resorts, whereas the unspoilt Costa Daurada and Costa del Camp de Tarragona were highly suitable for the establishment of camp-sites.

Timing as a Factor in Development

The importance of early local initiative for the subsequent development of international tourism has already been pointed out (Cals, 1982). Marketing strategies encouraged further concentrations of tourist accommodation without bearing in mind optimum limits of urban growth, beach capacity and environmental deterioration in general. Hotel and apartment accom-modation for short-stay holiday-makers, as a consequence, is concentrated in a relatively small number of large resorts (Fig. 7.6 and Tables 7.2 and 7.3).

The distribution of second homes is much more ubiquitous (Fig. 7.6 and Tables 7.2 and 7.3). Until the 1960s, the number of second homes was very limited, and these were mostly located in already established coastal towns, where minimum amenity and service requirements were available and rail connections with Barcelona generally existed. Second homes were therefore concentrated in the towns on the Costa del Maresme as far north as Blanes. The main exceptions were S'Agaró, an exclusive property development com-posed of detached mansions on a flat but rocky promontory on the Costa Brava between Sant Feliu de Guixols and Platja d'Aro, and Sitges, 40 km

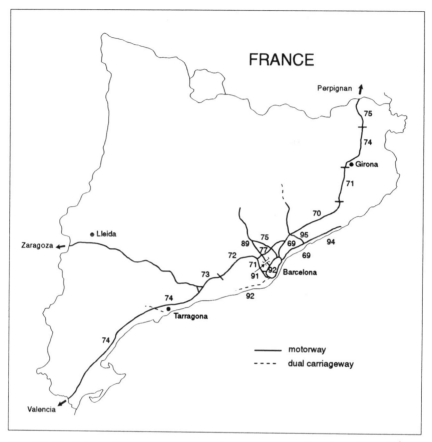

Fig. 7.5. Evolution of motorway construction in the coastal region of Catalonia.

southwest of Barcelona, where the flat land adjoining the beach to the west of the town was also occupied by a similar property development.

Second-home development after 1970 was much more land-consuming, not only because of quantity but also because of its morphology. The provision of public services and amenities outside the already built-up areas, together with the widespread use of private cars, encouraged the establishment of new locations and low-density developments. Municipal authorities permitted the construction of numerous property developments, known as *urbanizaciones*, in which investors provided sewage, mains water, gas and electricity, paved roads, street lighting and sometimes community leisure facilities as well as housing. As a result, hill slopes in pine copses or with a sea view, surrounding existing towns or along the coast were dotted with

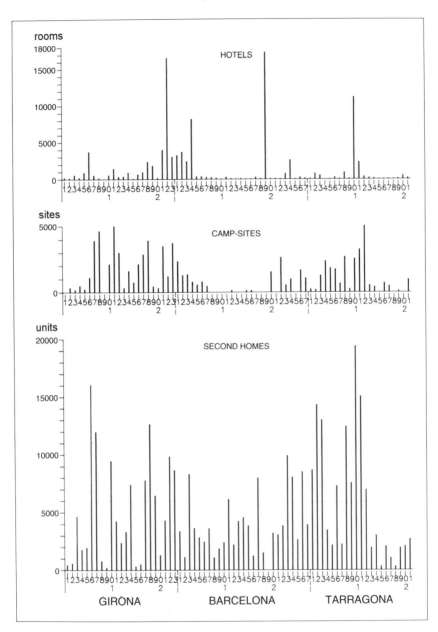

Fig. 7.6. Tourist accommodation in Catalan coastal municipalities (based on INE, 1992; Generalitat de Catalunya, 1994a, b).

Table 7.2. Tourism accommodation in coastal counties – *comarcas* – in Catalonia, 1992 (figures expressed in % of total capacity in coastal *comarcas*) (from Generalitat de Catalunya, 1993).

Comarca	Hotels	Camp-sites	Private
Alt Empordà (Port Bou–L'Escala)	8.5	**14.9**	**14.2**
Baix Empordà (Torroella de Montgrí–Santa Cristina d'Aro)	8.8	**22.4**	**15.3**
Selva (Tossa de Mar–Blanes)	**24.3**	8.6	8.3
Maresme (Malgrat de Mar–Mongat)	**18.4**	9.4	**12.7**
Barcelonès (Badalona–Barcelona)	**17.9**	–	0.2
Baix Llobregat (El Prat de Llobregat–Castelldefels)	1.5	9.8	6.5
Garraf (Sitges–Cubelles)	3.0	4.0	6.5
Baix Penedès (Cunit–El Vendrell)	1.6	1.7	**11.3**
Tarragonès (Roda de Barà–Salou)	**11.3**	**15.6**	**13.5**
Baix Camp (Cambrils–Vandellòs i L'Hospitalet de L'Infant)	3.2	**10.3**	7.1
Baix Ebre (L'Ametlla de Mar–Deltebre)	0.5	1.8	2.3
Montsià (Sant Jaume d'Enveja–Alcanar)	1.1	1.5	2.1
Total capacity (coastal *comarcas*)	193,816	218,088	1,434,902
% of total capacity in Catalonia	84.4%	88.3%	86.2%

low-density housing, mainly consisting of detached houses. This form of development was most common on the Maresme and Brava coasts. Development on the flatter sea fronts on the Costa Daurada and on the Costa del Camp de Tarragona generally consisted of low-rise, low-density and reasonably priced apartment blocks. Much of this development dates from as late as the 1980s, especially in Pals, Begur, Tossa de Mar on the Costa Brava (Fortià, 1993) and Vandellòs, Torredembarra, Creixell on the Costa Daurada (Áreas de Geografía, 1990).

Administrative Structure and Development

From this analysis it is obvious that the pattern of development emerged as different types of demand became evident. However, it was the administrative structure which made this spontaneous development possible. The coastline is highly fragmented in administrative terms, for a total of 69 municipalities are located along it (Fig. 7.1). No overall regional development plans have ever existed. As a consequence, spatial planning has been largely the responsibility of municipal authorities. Urban Development Plans – *Plan*

Table 7.3. Tourism accommodation in the principal coastal resorts in Catalonia, 1992 (figures are expressed in % of total capacity in Catalonia, and those below 1.0% are excluded) (from Generalitat de Catalunya, 1993).

Municipality	Hotel	Camp-site	Private
Roses	3.2	1.2	4.1
Castelló d'Empúries	–	–	2.7
L'Escala	–	2.0	2.4
L'Estartit	1.0	4.5	–
Pals	–	3.0	–
Palafrugell	–	–	1.9
Palamos	–	2.1	–
Calonge	–	–	2.0
Platja d'Aro	2.1	4.1	3.3
Sant Feliu de Guíxols	–	–	1.7
Tossa de Mar	3.3	2.5	–
Lloret de Mar	**13.7**	1.9	2.5
Blanes	2.4	3.5	2.2
Malgrat de Mar	2.4	2.0	–
Santa Susana	3.0	1.3	–
Pineda de Mar	1.9	1.3	2.1
Calella	**6.6**	–	–
Barcelona	**14.9**	–	–
Castelldefels	–	–	2.5
Sitges	–	–	2.1
Vilanova i la Geltrú	–	–	2.0
Cunit	–	–	2.2
Calafell	–	–	3.7
Tarragona	–	2.9	–
Vila-seca	–	–	1.8
Salou	**8.4**	2.9	**5.0**
Cambrils	1.9	3.5	3.9
Mont-roig del Camp	–	–	1.7

General de Ordenación Municipal – existed but were seldom rigorously implemented. These development plans were updated in the early 1980s, but their impact was very limited as they came after major tourist development and tended to legalize the status quo, making only minor modifications. In addition, during the period of rapid expansion, municipal revenues were totally insufficient to finance the provision of services and infrastructures. It was only after 1977 that such revenues gradually increased, although by that time the deficiencies were considerable. In a framework of legislative vacuum, administrative fragmentation and lack of public funding, it is easily understandable that private investment from diverse sources and for almost any purpose was warmly welcomed in most municipalities.

True municipal development policies began to emerge during the 1980s, as democratic structures took a firm hold. The new urban master plans were, in some respects, an expression of policy, even taking into consideration the restrictions imposed by previous development. In the present decade, policy definition has been taken a step further, as some municipalities have designed strategic plans. These include Sant Feliu de Guíxols, Sitges and Calafell.

Impact on the Natural Environment

The present pattern of tourism and recreation development on the Catalan coast is an intricate mosaic of long-established, compact villages and towns, engulfed in some cases by lower-density hotel and residential complexes; sprawling ribbon development of apartment blocks and terrace housing; *urbanizaciones* of detached houses perched on hill and cliff slopes; and short stretches of coast in their natural state or occupied by pine-covered camp-sites (García Olagorta, 1995). The different natural elements which comprise the coastal fringe reinforce the image of diversity and unequal development (Fig. 7.2 above).

Such development has obviously led to an unequal pattern of land occupation, which in turn has had unequal repercussions on the environment. For example, in the municipality of Lloret de Mar, 60% of the surface has been built up, whereas in neighbouring Tossa de Mar only 25% has been designated until the present for construction. Moreover, as a general rule little attention has been paid to landscape impact. On the contrary, sites with a view have been prime locations, especially for second homes. A web of houses spread over most inland slopes facing the sea and on rocky promontories. With few exceptions, only the most inaccessible sites have been preserved in their natural state. On the Costa del Camp de Tarragona, the four existing promontories – Roc de Sant Gaietà, Cap Gros, Punt de la Mora and Cap de Salou – have all been spoilt by the construction of residences and hotels.

Humid zones in particular have suffered considerable losses. On the southern Catalan coast, from Cubelles to Vandellòs, the systematic occupation for tourist and recreation purposes of almost the entire coastal fringe over the last 25 years has virtually eradicated these zones (Oliveras and Roquer, 1990). On the Costa Brava, part of the wetlands was occupied by an *urbanización* called Empuriabrava in the municipality of Castelló d'Empúr-ies, one of the earliest large-scale projects which involved the conversion of the lagoons into canals to permit access by boat to private residences. An adjoining sector of the wetlands has been preserved and been given the status of a Natural Park – Aiguamolls de l'Empordà. Other humid zones are, however, in danger of being occupied. These include the nearby Fluvià River mouth and the Baix Ter area on the southern limit of the Costa Brava (Fortià, 1993) and the Llobregat River delta lagoons (Folch *et al.*, 1989).

The provision of accommodation has been accompanied by the development of an extensive road network (Fig. 7.5). This is obviously a necessity, especially as both national and international demand depends largely on road transport for access to destinations. In fact, 60% of foreign tourists arrive in Catalonia by road (Generalitat de Catalunya, 1993) and, indeed, central government investment throughout the Franco era – up until 1975 – was primarily aimed at improving accessibility. On the other hand, the provision of most other infrastructures lagged far behind requirements, with serious negative consequences for the environment. This is especially true in the case of liquid waste disposal measures. Sewage treatment plants were not constructed along with accommodation, and for many years most waste was simply ejected into the sea. As occupation levels on the coast grew, so too did sea and beach contamination levels.

In recent years, numerous marinas and harbours have been constructed along the coast. In 1954 there were only 16 harbours, including the major ports of Barcelona and Tarragona, and they were used largely for commercial purposes and/or by fishing fleets (Bas *et al.*, 1955). By 1993 the number of harbours and marinas had risen to 44, with a further six scheduled for construction – an average of one harbour every 12 km – and a total of more than 20,000 moorings. These have further contributed to increase landscape impact, disrupt the natural process of beach formation and cause water pollution from petroleum products.

Resulting Conflicts and Problems

International tourism demand on the Catalan coast has suffered periodical setbacks since 1974, although it has continued to grow at a slower and more irregular rate. However, by that date regional demand for leisure facilities was increasing rapidly, with the result that total demand continued its fast upward trend. The spatial coincidence of these two parallel phenomena is, in fact, a source of conflict. Both sectors compete for land for construction purposes, for use of the roads, for space on the beach, etc. As a result, some resorts are oversized and optimum land occupation levels have been exceeded to such an extent that not only the natural but also the built environment is highly degraded. In some cases, competition for a spot on the beach is such that a small sector of it is usually cordoned off for the exclusive use of the members of the 'beach club', normally made up of second-home owners.

Administrative fragmentation is also a source of conflict. In some cases, there is rivalry between municipalities to become the largest, most popular or highest-quality resort on a certain stretch of coast. This has led to the drawing up of master plans designed to encourage development, as in the case of Blanes, where the pine copses bordering the beach to the west of the town and the River Ter delta wetlands were designated for construction. Such philosophies also underlie the construction of the majority of the marinas, considered a means of upgrading the entire municipality, as the possession of a boat – especially a cabin cruiser – is a prestige symbol among second-home owners. At the other end of the scale, municipalities which attempt to apply conservationist policies find difficulties, as they are engulfed by neighbouring developments. Water pollution does not respect municipal boundaries, and marina developments affect beach formation and renovation further along the coast.

The problems resulting from excessive, unplanned and relatively uncontrolled development are well known, and those relevant to this case have already been cited. However, an additional problem further aggravates those of degradation of the natural and of the built environments, which is obviously at least partly a consequence of these. Statistics clearly reveal that the Catalan tourist industry, like the entire Spanish industry, has encountered certain problems of a relative loss of demand and economic viability in recent years. It is not the purpose of this chapter to analyse these problems or their causes, but there is no doubt that one of the factors which have contributed to them has been the deterioration of the region's good image and this is certainly related to environmental quality.

The Relevance of the Sustainability Concept to This Case

The spectacular scenery and unspoilt coves of the Costa Brava and kilometres of deserted beaches on the Costa Daurada, which were the original source of attraction for tourists to Catalonia, cannot be restored. Development introduced not only transformations but also environmental degradation. In such circumstances, where the natural environment has suffered such profound modifications, one might question whether or not sustainability is a relevant issue. Certainly, sustainability defined as 'maintaining a balance between the use and conservation of the natural environment' is obviously not a viable option, except in certain isolated cases where small stretches of unspoilt coastline remain.

Other types of sustainability can, however, be sought. In the first place, economic stability can become an objective. In other words, the level, type and quality of supply should be adjusted to match foreseeable demand and carrying capacity. In the case of highly developed resorts this usually involves a reduction in accommodation supply, removing in particular low-quality and obsolete units which offer a low level of amenity provision. As a result total visitor levels tend to drop, thus reducing pressure on services and on the environment in general. There is, therefore, a clear link between environmental quality, albeit in a modified state, and economic viability in this concept of sustainability.

A second type of sustainability can also be sought, involving basically the maintenance of a high-quality built environment combined with certain essential vestiges of the former natural environment – a rocky promontory, a beach, a pine copse, for example. A new image has to be created. The deserted beach has disappeared for ever, but it must be replaced by an attractive built environment. In this respect, the expectancies of mass tourists have changed. They are, essentially, big city dwellers, and feel comfortable in an urban environment when on holiday. They do, however, expect similar standards in urban services and amenity provision and, if possible, a more relaxing and attractive atmosphere and a more picturesque setting – less congestion, lower noise levels, more flowers, trees and open spaces and more places to take a stroll. The issue in this case is, therefore, the sustainability of this new tourist product and new image.

Solutions Adopted in Catalan Tourism Resorts

In the face of growing difficulties to maintain economic viability, various solutions have been adopted. These measures have been implemented at three scales: national, regional and local levels. At the national scale, legislative measures have contributed to restrain or even reverse impacts. The Shore Act – *Ley de Costas* – passed in 1988 (Ley 22/1988), achieved the obligatory demolition of all buildings that had been constructed on the beach, and asserted some control over the future use made of a 150-m-wide coastal fringe. It obviously came too late to reverse the process of land occupation, and in some cases entered in direct contradiction with municipal master plans (Brau, 1993). For example, the sea-front buildings in some existing villages, such as Platja d'Aro and the fishermen's quarter at Torredembarra, had direct access to the beach without even a separating footpath. The solution adopted in both cases – one of those contemplated in the Shore Act – was the construction of a wide promenade on the beach, so that the front line of buildings was no longer on the beach!

At the regional scale, the designation of Protected Areas (Llei del Parlament de Catalunya 12/1985) has achieved some control of ribbon development, as in the case of the Parc Natural dels Aiguamolls de l'Empordà, already mentioned, which was created in 1983. However, the major problem which faced the entire coast was the problem of pollution and waste disposal. The Catalan *Generalitat*, the regional autonomous government, introduced a sanitation plan – *Pla de Sanejament* – in 1984 (Generalitat de Catalunya, 1984), mainly to eliminate the contamination of the sea by rivers. It is, nevertheless, at local level that the problem of adequate sewage disposal is being solved, but on a piecemeal basis. Various groups of neighbouring municipalities have reached agreement on the provision and financing of waste treatment plants, but no overall plan for the entire coast exists.

The Marina Plan – *Pla de Ports Esportius de Catalunya* – constitutes a second measure introduced at the regional scale by the *Generalitat* in 1983 (Gelonch and Carreras, 1983). It was a first attempt to analyse the coast as a single unit, and determined the most suitable zones for locating marinas and those where they should be avoided. Nevertheless, some of the many new marinas which have been built since 1983 are located at places re-commended for protection in the Marina Plan.

Certain problems are common to all tourist resorts, regardless of their size and the origin and type of demand and, in fact, it is at local – municipal – level that the majority of solutions are being found and implemented, which is hardly surprising as planning for the future is a municipal responsibility. Apart from the provision of public services – such as sewage and water-supply – already mentioned, there are two main areas of action: the urban centre and the waterfront. The conservation of traditional buildings, the creation of pedestrian precincts in the zones where commercial and leisure attractions predominate, the planting of trees and location of street furniture, and the paying of greater attention to cleanliness and the general appear-ance of the streets are all measures which have contributed to make town centres less congested and more user-friendly. On all waterfronts, the cleanli-ness of the sand, the provision of basic services – showers, deckchairs, etc. – refuse collection and the reduction of water pollution are common priorities. Specialized amenity provision – refreshments, water and beach sports and games, etc. – varies from one beach to another, as do the amenities in the vicinity. A few beaches are backed by dunes and pine copses, but where buildings and access roads were allowed to encroach on these natural forma-tions, pedestrian promenades and cycle tracks have been laid in most cases, in an attempt to integrate the beach and its surroundings into a single leis-ure area. Vehicle traffic is often discouraged or prohibited, and car parks are provided elsewhere.

A basic distinction should be made between the type of solutions implemented in areas where second homes predominate and those intended to satisfy the international tourism market. In the former case, the provision of complementary amenities in the form of golf-courses and/or marinas has been the common denominator in higher-quality resorts while, in lower-quality resorts, improvements have, by and large, been limited to basic ser-vices and amenities. In the case of international tourism, attention has been paid to accommodation supply – a better quality/price ratio in hotels, and a general upgrading of camp-sites – together with the diversification of the supply of complementary activities.

Although such initiatives are the result of a growing awareness of the importance of environmental issues in tourism development, they also reflect an increasing consciousness of the importance of environmental conservation in general on the part of the public as well as the authorities. Proof of this was offered recently in Palamòs, when a referendum was held to determine the future use of the only unspoilt beach in the municipality and, indeed, on a long stretch of coastline. The population voted against the granting of planning permission for a large property development and in favour of the area's conservation in its natural state. Legal procedures by the municipal authorities to have the area registered as a protected natural area are envisaged.

Conclusions

Environmental degradation cannot be pinpointed as the cause of the problems facing the Catalan tourist industry. Other factors have been identified, including the lack of flexibility of the tourism product; excessive control of the market by foreign agents; infrastructural deficits; and price rises which were not accompanied by a parallel rise in the quality of the product (Vera Rebollo, 1993). Nevertheless, the environment is heavily implicated in these factors, because defective infrastructures – poor sewage, for example – have negative repercussions on the environment, and an evaluation of quality implies not only the standard of accommodation, services and amenities but also environmental quality.

There is no doubt that a major consequence of large-scale tourism development in Catalonia, as elsewhere, has been the profound transformation of the environment. The tourism product now marketed is not the same as the original product, then based on an unspoilt natural environment and low frequentation levels. Attractiveness brings popularity and this in turn brings development. The difficulty is always how to reconcile development with environmental conservation. In the case of Catalonia, the former has obviously taken priority over the latter. The Catalan coast is the paradigm of rampant, uncontrolled development, and certainly demonstrates the worst contradictions between the classic model of mass tourism development and the environment, aggravated enormously in this case by the superimposition of international mass tourism with domestic residential tourism.

The principal lesson to be learnt is the necessity for planning: planning for development, planning for conservation and planning for sustainability. Special attention should be paid to the narrow coastal fringe – the beach and its immediate surroundings – for it is the most vulnerable, the most difficult to regenerate, the most unique and the most valuable tourist attraction. It is particularly important to avoid ribbon development, designating some areas for development and others for total protection. Overall planning at regional level is necessary, at least to regulate and coordinate local planning.

Moreover, optimum development levels should be defined at the outset. Overdevelopment endangers economic viability as well as the environment. This is clearly illustrated in Catalonia, where the Vilaseca/Salou/Cambrils coast, with a permanent population of 32,500 inhabitants and a summer population of over 300,000 inhabitants, has been adversely affected by market trends in recent years, whereas smaller resorts, such as Tossa de Mar, which depend largely on repeat visitors, have maintained their occupation levels and economic benefits constantly throughout the period. Finally, the provision of adequate infrastructures, in order not only to guarantee the high quality of the product but also to minimize environmental degradation, is a basic requisite.

A distinction must be made between large resorts and small or undeveloped places. Large resorts require high-quality – understood in terms of value for money – service and amenity provision and marked product differentiation, in order to enable them to cater for a clearly defined and durable market. Sustainability is, therefore, defined in economic terms, by sustaining a predetermined and constant demand – with adequate profit margins – on a long-term basis. Small resorts, on the contrary, should be encouraged to remain small, and avoid large-scale amenity provision, with the exception of basic services – related essentially to hygiene – concentrating on measures designed to protect the natural environment. This will enable planners to establish a balance between visitor levels and environmental conservation, and thus achieve a purer form of sustainability – alas, no longer feasible in highly developed resorts.

References

Áreas de Geografía (1990) *Atlas Ambiental del Litoral de Tarragona.* CICYT, Tarragona, 47 pp.

Bas, C., Morales, E. and Rubio, M. (1955) *La Pesca en España: I – Cataluña.* CSIC, Barcelona, 468 pp.

Brau, Ll. (1993) La planificació urbanística en relació amb la Llei de costes. In: *Estudis i Monografies 15, Perspectives del Medi Ambient als Municipis del Litoral.* Diputació de Barcelona, Servei del Medi Ambient, pp. 55–66.

Cals, J. (1982) *La Costa Brava i el Turisme: Estudis sobre la Política Turística, el Territori i l'Hoteleria.* Kapel, Barcelona, 271 pp.

Folch, R., Masalles, R.M., Miracle, R. and Bech, J. (1989) El delta i els estanys litorals del Llobregat, Barcelona. *Metropolis Mediterrània* 12, 84–86.

Fortià, R. (1993) La plana i el litoral de l'Empordà. In: Forità, R. (ed.) *El Medi Natural a les Comarques Gironines: l'Estat de la Qüestió.* Diputació de Girona, Girona, pp. 203–281.

García Olagorta, G. (ed.) (1995) *Aeroguía del Litoral de Catalunya.* Planeta, Barcelona, 256 pp.

Gelonch, G. and Carreras, J.M. (eds) (1983) *Plà de Ports Esportius de Catalunya*, 3 vols. Direcció General de Ports i Costes, Generalitat de Catalunya, Barcelona.

Generalitat de Catalunya (1984) *Pla de Sanejament de Catalunya*, 3 vols. Generalitat de Catalunya, Departament de Política Territorial i Obres Públiques, Barcelona.

Generalitat de Catalunya (1993) *La Temporada Turística a Catalunya 1992*. Generalitat de Catalunya, Departament de Comerç, Consum i Turisme, Barcelona, 177 pp.

Generalitat de Catalunya (1994a) *Catalunya: Hotels '94*. Generalitat de Catalunya, Departament de Comerç, Consumi Turisme, Direcció General de Turisme, 245 pp.

Generalitat de Catalunya (1994b) *Catalunya: Campings '94*. Generalitat de Catalunya, Departament de Comerç, Consumi Turisme, Direcció General de Turisme, 36 pp.

INE (Instituto Nacional de Estadística) (1992) *Censo de Viviendas 1991: Avance de Resultados*. Instituto Nacional de Estadística, Madrid, 198 pp.

Ley 22/1988 de 28 de julio de Costas. Real Decreto 1471/1989 de 1 de diciembre (which established the general regulations for the development and implementation of the Ley de Costas).

Llei del Parlament de Catalunya 12/1985, de 13 de juny, d'espais naturals (in which Chapter III created and regulated a Plan for Natural Areas). Published in: Generalitat de Catalunya (1992) *Pla d'Espais d'Interès Natural*, 5 vols. Generalitat de Catalunya, Departament de Política Territorial i Obres Públiques, Departament d'Agricultura, Ramaderia i Pesca, Barcelona.

MCT (Ministerio de Comercio y Turismo, Subdirección General de Planificación y Prospectiva Turística-Estadística) (1994) *Anuario de Estadísticas del Turismo 1993*. Ministerio de Comercio y Turismo, Secretaría General de Turismo – TURESPAÑA, Dirección General de Política Turística, Madrid, 484 pp.

Miguelsanz, A. (ed.) (1983) *Llibre Blanc del Turisme a Catalunya*. Generalitat de Catalunya, Departament de la Presidència, Barcelona, 382 pp.

Oliveras, J. and Roquer, S. (1990) Le littoral méridional de la Catalogne. Agriculture, tourisme, industrie: un partage difficile de l'espace. In: Casa de Velázquez (ed.) *Géographie d'une Espagne en Mutation. Prospections Aériennes II*. Casa de Velázquez, Madrid, pp. 53–72.

Vera Rebollo, F. (1993) Actividades y espacios turísticos. In: Méndez, R. and Molinero, F. (eds) *Geografía de España*. Ariel, Barcelona, pp. 468–503.

8 Decline of a Mediterranean Tourist Area and Restructuring Strategies: *the Valencian Region*

FERNANDO VERA AND REBECCA RIPPIN

Introduction

A large number of tourist developments built in the Mediterranean region during the 1960s and 1970s to accommodate mass tourism are currently having to face the challenge of the emergence of alternative destinations, the diversification of motivations in demand, and threatening decline and even survival. Many have responded by taking steps to improve the competitiveness and sustainability of the tourist product on offer. The Valencian Region of Spain, where tourism plays a fundamental role in the regional economy, exemplifies the situation described above. Resorts specializing in mass tourism commercialized by foreign tour operators – including Benidorm, a classic and, indeed, flourishing destination – have seen the emergence nearby of more élite resorts where residential tourism predominates, a product based on the sale or rent of accommodation integrated within residential complexes, some of which constitute colonies of north European nationals.

The overall interpretation of these developments in a mass tourism market supports the theory of the evolution of tourist areas based on the life-cycle concept (Butler, 1980). Following this theory, the relationship between the market and the product is used to model the dynamics of a tourist resort from its initial shaping throughout its later development, thus providing a tool with which to determine the limitations and define the inflexible nature of such resorts. Various studies have analysed this classic tourist model: in Britain (Urry, 1990); the Valencian Region and, more specifically, the case of the Province of Alicante (Pedreño Muñoz, 1986); and the Costa del Sol, Andalusia (Torres Bernier, 1989).

However, it was Agarwal (1994) who, with sound arguments, went beyond Butler's basic resort life-cycle concept to describe in detail the alternatives available when faced with the stagnation stage of a tourist model. Pedreño Muñoz (1986), Aguiló and Torres (1990) and Aguiló (1992) all make interesting contributions to the analysis of examples of situations of saturation in the Spanish context. The case-studies they present serve to illustrate the process of tourism development, as well as highlighting difficulties resulting from its unstructured and untidy development. These difficulties are analysed in Vera Rebollo and Marchena (1990) from a territorial perspective and in Vera Rebollo (1990, 1992) within a Mediterranean framework. Favourable conditions can lead to successful tourist seasons – such as the summer of 1994, a record year in terms of tourism revenue in the Valencian Region, given the weakness of the local currency and political and social instability in potential competing destinations, which only a few years earlier had posed a threat to resorts such as Benidorm. Nevertheless, it is widely recognized that structural problems, such as those highlighted above, hinder the consolidation of sustainable tourism development in the Mediterranean arc and similar tourist areas in Spain.

In an attempt to design strategies for the future development of a sustainable model of tourism, this chapter analyses some of the policy inititatives proposed to counter the fall in tourist demand and upgrade the product. An analysis is made of an ambitious study commissioned by the Autonomous Regional Government of Valencia, which was intended to provide a Master Plan for Tourist Areas – *Plan Director de los Espacios Turísticos* – (FCAET, 1995), on which future tourism policy in the region was to be based. This plan, together with other initiatives, constitutes a creative response to circumstances which, at present, threaten the viability, and hence sustainability, of the Valencian tourist industry.

Structural Problems in Tourism Destinations in the Valencian Region

Tourism development without urban planning creates the very problems it later faces when tourism demand no longer finds it appealing. Areas of highest tourist density and most widespread land occupation for tourism- and leisure-related activities and services are now facing the consequences of this type of development.

A closer examination of the characteristics of three decades of tourism development in the Valencian Region shows that, in terms of structure and natural advantages, the pattern of development in the region is similar to simultaneous processes in the Balearic Islands, the Canary Islands, Andalusia and Catalonia. However, in the case of the Valencian

Region, a general failure to offer a diversity of tourist products has had particularly negative results, especially in the face of market changes, such as: variations in consumer tastes, exchange rates, seasonality, successful marketing campaigns made by competing destinations, etc. Whilst many Spanish tourism and leisure areas have failed to address the issue of the diversification of risk in tourism, this failure is even more marked in the Valencian Region, since, in contrast to other nearby destinations, it has not been able to combine its sun/sand/sea product with other alternatives, such as winter sports, rural, landscape or historic heritage tourism, sport or congresses.

The existing product – massified and undiversified – is unable to satisfy current demand, which is increasingly more discerning. There is a need for a wider range of attractions to be made available in areas which have, as yet, failed to explore – much less, promote – their potential, and for new alternatives based on a new consumer culture that is increasingly influenced by the awareness of issues such as environmental quality and natural resource conservation. The emergence of new market sectors illustrates the importance of planning tourism from a territorial point of view. Land must be regarded as a resource in itself and not merely as a setting within which to locate tourism, as has tended to happen in the past. Therefore, strategies aimed at consolidating existing, or shaping new, tourist areas in the future must include regional planning.

It should be stressed at this point that the introduction of new strategies aimed at upgrading supply – through product recycling and the improvement of the appearance of saturated areas – and diversification – through the creation of new, complementary products – does not necessarily imply the rejection of the star product of the Spanish tourism industry, but rather provides an opportunity to reconvert low-quality accommodation and abandon the trend towards the continuous growth of supply. Emphasis should therefore be placed on product improvement to enable resorts to compete with alternative destinations on the basis of quality. This message must be understood by professionals at all levels in the industry and public administration.

The recent lifting of air transport restrictions within the European Union and the possible end to conflict in current war-affected areas in the Mediterranean are factors which may lead to the discovery or revival of alternative holiday destinations. Within this context, established resorts, such as those along the Valencian coast, should use their traditionally most successful product as a basis for the development of new strategies, given that the importance of specialization and high quality are beginning to be recognized. Competitive prices – an essential advantage for any resort in its transition period – must be regarded at this stage as an unstable and temporarily differentiating factor which should not be depended upon for success on a long-term basis. Changes are vital if tourism is to be valued as a real factor in the sustainable development of the region.

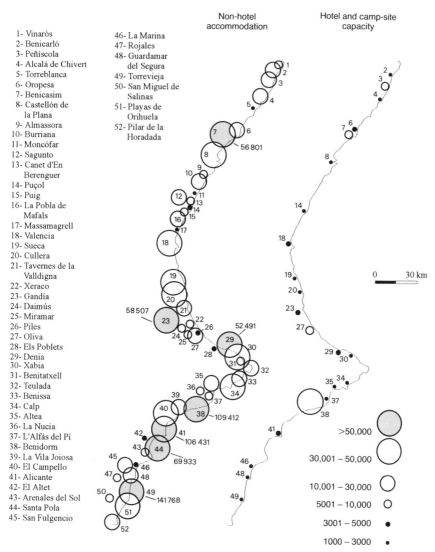

Fig. 8.1. Tourist accommodation on the Valencian coast, 1994 (based on FCAET, 1995).

Strategies in Tourism Policy Design for Consolidated Destinations

Constraints and limitations

The detailed analysis of a case-study makes it possible to identify the constraints and limitations which prevent a region from improving and sustaining its capacity to be competitive. Generally speaking, these affect both the quality of the product and its suitability to satisfy demand.

In the first place, measures taken by the public sector in an attempt to improve competitiveness tend to disregard the wider spatial context and the need for regional integration. If future tourism policies are to be effective, they require a degree of commitment from the different sectors of the public administration. Past experience has shown that, all too often, public-sector actions have tended to be partial, fragmented or even absent, and there are clear examples of a lack of coordination between different state departments or ministries and between national, regional and local administrative bodies. In the case of the Valencian Region, the *Plan Director de los Espacios Turísticos* attempts to provide a framework and an instrument for future development through administrative consensus and coordination.

Secondly, tourism policies have tended to concentrate on product marketing and promotion, while product quality has been neglected or, even worse, target markets have not been properly identified. Clearly, marketing must be seen as a basic tool which should complement, but never replace, the process of salvaging and reorientating a floundering tourist product. Such a process should lead to product differentiation and an improvement in the quality of the tourist industry's infrastructures, one of the recognizable weak points of the Valencian product and indeed of many other Spanish tourist areas. The ultimate objective is to recover a prominent position among leading destinations, justifying public- and private-sector investment by making it profitable and thus achieving a form of sustainable development.

Thirdly, investment from the private sector is not always forthcoming. In the Valencian Region, investment and entrepreneurial know-how are lacking, as a result of a failure to fully appreciate the importance of tourism, as indicated by the fact that a large proportion of the revenue earned through this activity is siphoned off towards other sectors of the region's economy. Not surprisingly, then, many hotels are unable to carry out reforms and set up complementary activities. Private initiatives therefore play an essential role in the entire process of general reform and improvement of the tourism industry. Parallel to public sector actions, these will together transform and upgrade the entire tourism environment.

At present, therefore, the product's positioning with respect to new trends in demand is weak. These same trends will, however, serve to force a dynamic reaction within the whole tourist system.

Strategies towards competitiveness

Although it is important to stress strategies related to marketing and market actions, those which affect the product itself and aim at enhancing its quality should also be considered. To ensure tourism specialization in the Valencian Region and, as a result, its competitiveness, attention should be centred on unexplored or insufficiently developed possibilities (Vera and Monfort, 1994). These include the following.

- Actions aimed at the proper identification and description of the product. It is important to stress the benefits to be obtained from successful product promotion, which requires well-structured professional marketing campaigns in order to increase market penetration of a product which is already known. It should be stressed that the differentiating factor in the tourist product is real competitive advantages rather than traditional, and often saturated, comparative advantages – especially those related to climate and low prices. In this respect, public-sector involvement in marketing and promotion should be less exclusive; there should be room for partial decentralization or privatization, thus offering greater participation to the specific decision-making bodies involved in management development and a more prominent role to the agents who intervene in the process of product communication.
- Actions related to distribution or the way in which the tourist product is made available to the consumer. This factor is particularly relevant in the Valencian Region, where the oligopoly of demand has existed in most resorts, since the take-off of tourism in the late 1960s, due to the strong control exercised by tour operators. To counter this situation, business associations, incorporating agents responsible for all types of holiday accommodation, should be set up with the aim of establishing their own distribution networks for the Valencian tourist product within the framework of existing opportunities and possibilities.
- Strategies to introduce instruments aimed at increasing consumer satisfaction levels, since the degree of satisfaction is the key to maintaining tourist demand and increasing market share. Throughout the development of tourism in Valencia, the implementation of quality controls of the product and the service offered to consumers has largely been absent. It is only recently that professionals and public administration departments have started to aim at achieving quality standards capable of satisfying demand, instead of concentrating their efforts on increasing visitor levels. Many instruments exist which can be used both to guarantee quality and to measure the level of consumer satisfaction: tourist information systems; centralized regional reservation networks; consumer rights bureaux; increasing awareness on the part of public administrations, entrepreneurs and citizens; clear signposting; environmental conservation, etc. All such measures point to the much-needed professionalization of tourist services and the consequent recognition that tourism is most definitely not a temporary or part-time industry, but one which requires adequate human resource training and cannot improvise on innovation.

A Territorial Strategy for Sustainable Tourism Development

Primary objectives

As has already been pointed out, tourism development in the majority of long-established destinations where the main attractions are sun and sea took place on a massive scale and in a disorderly fashion, as supply attempted to keep pace with growing demand in especially attractive natural environments. The resulting conflicts have also been identified, as have the basic strategic objectives of achieving greater stability in tourism and a higher-quality supply. The fundamental aim, therefore, must be to transform territorial planning and adapt tourist operations to conform to a spatial logic designed to reduce to a minimum environmental degradation associated with the provision of supply and its use. The polarization of initiatives so characteristic of previous growth stages should be avoided, and new methods of organization better adapted to the general process of territorial development and redistribution applied. Priority strategies capable of achieving spatial coherence within tourism should include the following.

- A territorial strategy which would provide a framework within which problems of saturation may be overcome, through the regeneration and restructuring of long-established resorts. Existing development in traditional destinations is reinforced, as their ability to adapt to new demands is enhanced through the promotion of new attractions, especially those which do not involve increased carrying capacity.
- Spatial logic which would facilitate the growth of alternative destinations within a region through the development of new tourism products. The first step would be to study the potential of less developed areas nearby which offer considerable scope for the development of alternative or complementary products. This would lead to the creation of new business opportunities, and the net result would be the controlled, sustainable development of such alternative tourism products.
- Territorial planning on a truly regional scale, since tourism is capable of contributing to restoring an equilibrium in the creation and social distribution of wealth through more efficient use of land. Given the complexity of the problems that exist, limitations of scale must be overcome if new opportunities are to be identified. New instruments within the sphere of public-sector intervention and ways of redirecting private initiatives are necessary. The ultimate goal must be to establish new forms of organization and patterns of land use (Quero and Leira, 1990). Intervention by the public sector should be orientated towards correcting the deficiencies detected in present tourist areas and facilitating their evolution within the framework of a general process of restructuring, whereby tourism adopts the role of a strategic activity as an instrument of regional and local development, within the context of the transformation of mass tourism to post tourism (Urry, 1990).

The key to achieving these objectives is innovation, rationalization and high quality. Two conditions must exist if these objectives are to be attained. Firstly, the administration must demonstrate its willingness and determination to provide and, indeed, impose effective coordination, thus helping to establish the basis on which sectoral initiatives may be articulated. Such a process exceeds the limits of urban planning at municipal level and thus

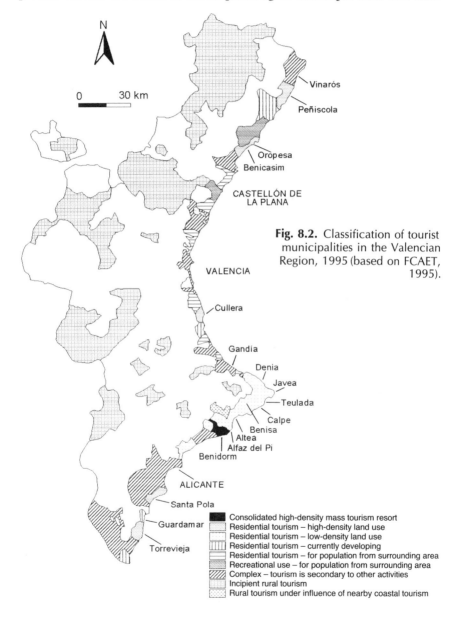

Fig. 8.2. Classification of tourist municipalities in the Valencian Region, 1995 (based on FCAET, 1995).

presupposes the involvement of regional government departments at all levels. Secondly, the fact that spatial planning has to take account of the environmental and natural dimension must be understood as a real step forward, and should be included in all policies aimed at regenerating consolidated destinations. Examples of territorial planning can be found in other countries and regions that have accepted the challenge of improving their competitive position in the tourist industry (Antón Clavé, 1993).

Territorial restructuring and regional integration

The weaknesses of the mere articulation of sectoral and physical planning as the basis of territorial planning of tourist areas become apparent as soon as signs of saturation and loss of competitiveness of destinations appear. At this stage, the contradictions inherent in a policy of continuous growth of supply become obvious: imbalance between supply and demand; high seasonality; a static and massified supply; and aesthetic and functional degradation. Only then is the introduction of alternative tourist products contemplated, and land identified as a resource in its own right, instead of being considered simply as a setting in which tourism is located. The identification of the processes and factors involved finally appears in response to an increasingly specialized demand.

Not surprisingly, the regions first affected by this change in attitude are those that are suffering the decline of their previously successful sun/sand/ sea resorts, e.g. the Mediterranean. The resort life-cycle theory (Butler, 1980) has been used not only to describe the relationship between products and markets, but also as a framework for establishing instruments to restructure the seasonal and spatial dimension of tourist resorts. As a result, restructuring proposals related to the life-cycle concept began to be included in analyses of European tourist resorts at a regional and local level (Antón Clavé, 1993) and in strategic planning towards sustainable forms of tourism, as in the case of the West Country in England (ETB, 1991). Solutions for the problem of loss of competitiveness in traditional Spanish resorts serving the mass sun/sea market are now being sought within the framework of regional territorial strategies, spurred on by the conviction that the very contradictions resulting from the pattern of construction of the tourist area should find their solution in actions aimed at planning and organizing elements derived from the new forms of economic organization in the area. Some unique geographical areas, such as Gran Canaria, have in fact already undergone a process of strategic territorial development. This represents a new approach to territorial policy, for tourism is regarded, for the first time, as a fundamental object of territorial transformation. Rigid models of physical planning out of touch with social and economic reality are rejected. This approach also moves away from the promotion of mass demand, indiscriminate consumption and the uncontrolled exploitation of the environment, and towards stimulating an awareness on the part of estate agents of how to develop and exploit tourist products. In fact, for the first time 'actions' are called 'products' and spatial or differentiating elements are discovered or identified and referred to.

It is, however, not enough to simply add alternative tourism products to a regenerated urban environment. The role played by the introduction of such elements in a leisure environment in producing new spatial forms in their immediate surroundings is also important. This can, in fact, lead to the formation of renewed products, which should include complementary facilities, especially those not directly linked to accommodation, and can, therefore, serve to upgrade the tourist area by adding to its commercial appeal. The option of introducing the conceptual principles of territorial planning of tourist areas also requires the reconsideration of the role played by natural and agricultural spaces within these areas. The implication is that diverse land uses can actually be complementary and can enhance the appeal of new tourist products.

The second sphere of action in global territorial policy is the development of new products in areas which are not at present important tourism destinations. Inland and mountain areas are especially affected by these developments, where 'green' tourism – hiking, hunting, adventure sports – and winter sports can be established. These forms of tourism are developing rapidly in Spain in, for example, Cantabria and Andalusia. Tourism related to cultural and architectural heritage is developing in, for example, Castilla y León, whilst urban tourism, linked not only to business and congresses but also to heritage, is expanding in many smaller cities as well as in the major cities of Madrid and Barcelona. Other forms of tourism – educational, social, health – complete the panorama of new products which are currently being introduced in Spain. The underlying objective is to integrate tourism within the more general objectives of overall regional development, in an attempt to encourage territorial and social rebalancing. Alternative forms of tourism, using existing resources and assets, offer considerable potential for future growth and constitute an important factor in the future development of tourism, from a territorial and environmental perspective.

Spanish government tourism policy

Intervention in tourism from a territorial perspective makes it possible to coordinate different types of tourism and a number of tourist resorts, and to articulate complementary areas and avoid territorial conflicts, while simultaneously enhancing the quality of the product and diversifying potential demand through segmentation. Such territorial considerations in tourism, aimed at optimizing resources and increasing competitiveness by upgrading consolidated destinations and encouraging the establishment of new assets, were also defined as objectives by the Spanish Ministry of Industry,

Commerce and Tourism in 1992 in its *Plan Marco de Competitividad*, also known as *Plan FUTURES* (Secretaría General de Turismo, 1992).

The Plan was elaborated jointly by the Ministry and the Autonomous Regional Governments, with the collaboration of the social agents involved in tourism development. In the context of changing domestic and international tourist markets, the Plan defines objectives aimed at modifying strategies in order to revitalize Spanish tourism and improve competitiveness. Emphasis is placed on the integration of traditional destination areas in overall regional structures by facilitating their adaptation to the changed circumstances, and on the creation of new products in less developed areas. The ultimate objectives of the Plan are of a social, economic and environmental nature.

Within the framework of the *Plan FUTURES*, specific programmes offering financial support for individual initiatives were established. The structural recomposition of consolidated destinations through the improvement of both the urban and natural surroundings was outlined in the *Plan de Excelencia Turística*. Some of the operative programmes contemplated involve specifically territorial actions, with the objective of achieving regional development, an aspect of special interest to the Autonomous Governments. As a result, financial support is offered for measures taken in inland tourist areas, and aimed at introducing new products in an attempt to benefit the least developed areas.

It was, in fact, the loss of competitiveness of Spanish tourism which led to widespread recognition of the important role of territory and environment, together with the need for spatial restructuring, in the quest for competitiveness. Product diversification and product upgrading have become key objectives within the overall objective of achieving the sustainability of a tourist area, defined in terms of commercial success, continuous demand and environmental protection. Where the sun/sea product is predominant, the environment is recognized as a key factor and, at the same time, as its weak point, since environmental quality in mass tourism resorts has suffered a decline due to the very nature of development. The *Plan FUTURES* is an exceptionally positive measure, which takes into account all three types of objectives – social, economic and environmental improvement.

Applications of the Plan FUTURES at a regional scale

In an attempt to achieve regional integration, territorial planning as a tourist policy strategy has been applied in Andalusia, where tourism was, in fact, identified as a primary factor in territorial rebalancing (Marchena, 1988; 1990). Regional strategies involve interventionist policies in tourism, and new actions attempt to apply spatial logic and to take account of environmental questions – analysis of physical conditions, integration of landscape values, architectural design. This shows the determination of the administrative bodies to coordinate initiatives in the field of tourism with those of other industrial sectors – environment, agriculture, transport and public works. The territorial model applied in Andalusia, designed to give spatial coherence to the tourist phenomenon, is presented in *Planes Territoriales del Litoral* and *Programas de Desarrollo Integral del Turismo Rural*, which together constitute the *Plan DIA* (Junta de Andalucía, 1993). On the one hand, traditional coastal tourist areas are encouraged to improve present facilities, to adapt to changing demand requirements and to become more competitive, by promoting new complementary activities without increasing accommodation capacity. On the other hand, the promotion of tourist development in inland areas is also contemplated in the Plan. Specific measures include: *Programa de Ordenación Turística del Territorio*, which specifies tourist land and redesigns the layout of consolidated areas; *Mejora Global de la Oferta*, which aims to improve the quality of the product on offer; *Plan Piloto de Recuperación de Zonas Saturadas*, to regenerate saturated areas; and *Plan de Mejora del Entorno*, to introduce improvements in the environment in general. The overall objective is to establish integrated planning of tourist destinations and their surrounding areas, paying particular attention to the quality of both urban and natural environments. As a result, it is hoped to improve returns – economic, social and environmental – on the use of the land for tourism.

The repercussions of improvements and reforms undertaken in some consolidated coastal resorts are far-reaching. Such is the case of Calvià in the Balearic Islands, where tourism has been the basic element on which restructuring and upgrading plans have focused. The net result is important qualitative improvements to the conventional product (Leira and Quero, 1988). Torremolinos is another example of this kind of resort, where environmental degradation was the outcome of the success of the sun/sea product, which lasted until the mid-1980s. The resort was then forced to face problems of loss of competitiveness, together with all the imbalances inherent in the traditional tourist model (Ayuntamiento de Torremolinos, 1994). Shock remedies were devised to counteract this situation. The underlying philosophy was that the introduction of urban improvements and environmental conservation measures, together with modernization, would generate new business opportunities. Measures designed to repair environmental deterioration in an attempt to recover lost ground in demand and competitiveness were adopted within the framework of *Plan FUTURES*; *Plan de Excelencia Turística* and *Plan DIA* initiatives were also incorporated into several of the specific projects. In Torremolinos, the General Urban Master Plan – *Plan General de Ordenación Urbana* – is also being revised. Specific projects include the widespread rehabilitation of buildings and the countryside, the restructuring of urban areas and the provision of new infrastructure. More general, long-term measures include the creation of a new territorial development model. The cost of the entire project is estimated at

1122.750 million pesetas over the 1994/95 period, 383.500 million pesetas of which will be financed by the private sector.

Resorts such as Torremolinos and Calvià, which have adopted the *FUTURES/Excelencia* guidelines as a basis for action, have increased their chances of receiving financial support from government funds by becoming, to all intents and purposes, showcases for current tourism policy in Spain. But official financial aid will be of little use if they do not win the support, enthusiasm and active participation of the sizeable business class in the task of upgrading their resort. Nevertheless, not all initiatives aimed at upgrading receive central government financial support. In fact, many examples of measures financed and carried out exclusively at municipal level are based on ambitious marketing and commercialization campaigns, e.g. in Benidorm. Such actions must, however, be accompanied by regional and urban restructuring, if the product itself is to be improved. This is the direction in which other tourist towns are moving, by applying strategic plans which combine initiatives related to the product as well as to its marketing. The control of supply and the development of new products form part of urban planning structures which are viewed as the means of introducing such measures.

One Application of the Territorial Model: the *Plan Director de los Espacios Turísticos* in the Valencian Region

Tourist areas within the Valencian Region have much in common with the resorts mentioned above and, indeed, recent policies which have been applied at a local and regional level reflect the recognition of the importance of introducing policies aimed at establishing conditions which will enhance their chances of survival in the modern tourist market, in other words, create sustainable tourism structures. The White Paper for Tourism in the Valencian Region (ITVA, 1990) highlighted as the prime objective of its tourist policy the maximization of the contribution of tourism to income and employment, thus identifying the strategic nature of tourism and of its territorial and social implications. The second objective outlined in the White Paper stressed the role of tourism as an instrument of economic and regional development, and highlighted its additional role as a potentially effective and efficient means of conserving environmental resources and cultural factors.

Following the publication of the White Paper, the decision was made to devise a Master Plan for Tourist Areas – *Plan Director de los Espacios Turísticos* (FCAET, 1995). The first step in designing this Plan was the identification of the spatial conflicts generated as a result of tourism development, and the recognition of the extreme fragility of the existing forms. This clearly demonstrated the need to enhance quality, diversify demand and encourage other kinds of tourism which could contribute to a process of territorial and socioeconomic rebalancing by spreading tourism more widely and thus relieving pressure on the overloaded – from both an urbanistic and a demographic point of view – coastal fringe. On the basis of this general diagnosis, various strategies were proposed which, as has occurred in other Mediterranean regions, affected both the product and its distribution. The basic aims are twofold:

- to maintain and promote tourism development in consolidated coastal areas, by creating new attractions and improving the competitive position;
- to diversify development by incorporating new inland areas which were formerly neglected in the process, but where both tourist assets and business opportunities exist.

The *Plan Director de los Espacios Turísticos*, therefore, focuses on the rational use of space and resources. It establishes the basis for coordination between different regional governmental departments through sectorial and territorial plans, reinforcing its supramunicipal and multidisciplinary approach. Thus an entirely new and comprehensive instrument for public intervention and coordination has been established, an instrument which will attempt to redirect private initiative towards the development of more rational organizational patterns and forms which take into account the carrying capacity of each area and the need to diversify and upgrade the product offered.

The Master Plan attempts, once and for all, to transmit the idea of rational, planned growth in a region where property operations have served to cover vast areas with a massified supply of non-differentiated accommodation. This situation illustrates the need to understand and deal with territorial problems on a larger scale, thus making it possible to identify real possibilities and limitations, and to establish typologies and delimit different areas of tourist supply by applying the concept of a tourist system (Clary, 1993). Such areal differentiation, through which large areas with similar problems can be identified, is a prerequisite for initiating an integrated planning process. In order to make the strategic plan operative, a framework of administrative agreement and cooperation has been established, consisting of global, orientative proposals which will guide long-term policy in drawing up institutional, economic and technical agreements. The Plan is intended to serve as as a point of reference for both public bodies and private development agents.

An additional aspect of the Plan is the provision of guidelines for the control and reversal of current degradation, especially in saturated coastal areas, and for the introduction of new developments. Recommendations are both of a spatial nature – diffusion towards new areas – and product-orientated – encouraging new forms – with the emphasis firmly placed on environmental protection and on local development, seen from a supramunicipal perspective. As a result, the measures proposed necessarily imply actions and policies involving different regional government departments, and will be implemented in two ways.

- The instruments foreseen in the *Ley de Ordenación del Territorio de la Comunidad Valenciana* (Ley 2/89), will be used to apply a series of planning measures in such a way that tourism strategies will form part of sectorial planning, thus ensuring coordination between regional and local government departments.
- The recently created Interdepartmental Tourism Commission of the Valencian Regional Government will serve to implement strictly sectorial actions which affect different government departments. Thus the viability of the Plan is guaranteed, even in the fundamental aspect of financial resources, since these will be provided by each of the different administrative sections involved. This system of coordination is unique in that it overcomes the usual lack of coordination between government departments of affairs related in some way to the tourism industry. The departments affected include Environment, Public Works and Transport, Employment, Industry and Health, among others.

The Plan recognizes the role of tourism as a strategic activity and the value of land as an exclusive resource. This favours the appearance and growth of new and more diversified tourist products, created and exploited according to new criteria and under different forms of organization. The implementation of the various specific projects resulting from the Plan will depend upon the capacity and uniqueness of individual tourist areas and on product/market strategies. Projects include:

- actions involving physical and ecological conservation – parks and nature reserves – and the promotion of natural assets – ecotourism, protected areas in urban planning;
- restoring heritage and historical buildings in rural and less accessible inland areas;
- regeneration of consolidated destinations through upgrading and the introduction of elements and attractions which improve the deteriorated image of certain urban tourist areas;
- innovative tourism operations to diversify supply and increase its quality. Specially privileged areas will be pinpointed for the introduction of new products;
- identification of functional areas, where mass supply generates suburbanization and dependence on traditionally central places, and encouragement of the growth of new functional centres;
- improvement of access between coastal and inland areas, especially in rural or mountainous areas in close proximity to the coast, where the expansion of tourism is feasible and the encouragement of its growth constitutes a complementary economic activity;

- regeneration of deteriorated natural and urban environments;
- plans for the provision of tourist facilities within designated territorial units in order to ensure the profitability of investment and the complementary nature of initiatives.

Final Comments

The guidelines of the Valencian *Plan Director de los Espacios Turísticos* establish two essential principles for the future development of tourism. The first is cooperation among all the agents and institutions involved in specific project design and implementation. The second is the need for tourism to be recognized as a strategic activity, and this can be achieved only if land is considered the key resource in tourism development. This constitutes a new way of understanding tourist areas within the context of sustainable development. The region should no longer be regarded simply as a setting in which tourist activities are located, but rather as a medium for their development. The limits of the capacity and potential of each region must be recognized in order to reveal its true role. Hopefully, a new era of tourism development is emerging, one in which tourism activity is valued more positively and evolves in consonance with more responsible management of the entire territory and natural resources. As a result, it will be possible to assess future potential more realistically, create new, more viable forms of tourism and revitalize existing ones. Such a step towards overall regional planning and cooperation is vital if regions where the economy depends largely on the tourism industry are to guarantee the sustainability of their tourism products.

References

Agarwal, S. (1994) The resort cycle revisited: implications for resorts. In: Cooper, C. and Lockwood, A. (eds) *Progress in Tourism, Recreation and Hospitality Management*, vol. 5. University of Surrey, Guildford, pp. 194–208.

Aguiló, E. (1992) La posición competitiva de las regiones turísticas mediterráneas españolas: posibilidades de la política turística. *Papers de Turisme* 8/9, 75–92.

Aguiló, E. and Torres, E. (1990) Realidad y perspectivas del sector turístico. *Papeles de Economía Española* 42, 292–305.

Antón Clavé, S. (1993) Consideraciones sobre la reordenación y revitalización de nucleos turísticos, revisión de procesos y experiencias. *Papers de Turisme* 11, 33–48.

Ayuntamiento de Torremolinos (1994) *Torremolinos: Plan FUTURES 1994–95.* Ayuntamiento de Torremolinos, Torremolinos, 71 pp.

Butler, R.W. (1980) The concept of a tourist area cycle of evolution: implications for management of resources. *Canadian Geographer* 24(1), 5–12.

Clary, D. (1993) *Le Tourisme dans l'Espace Français*. Masson, Paris, 358 pp.

ETB (English Tourist Board) (1991) *A Regional Tourist Strategy for the West Country: Summary, 1992–1996*. ETB, London, 92 pp.

FCAET (Fundación Cavanilles de Altos Estudios Turísticos) (1995) *Bases para la Redacción del Plan Director de los Espacios Turísticos de la Comunidad Valenciana*. FCAET, Alicante, 234 pp.

ITVA (Institut Turístic Valencià) (1990) *Libro Blanco del Turismo en la Comunidad Valenciana*. ITVA, Valencia, 394 pp.

Junta de Andalucía (1993) *Plan DIA: Plan de Desarrollo Integral del Turismo en Andalucía*. Turismo de Andalucía, SA, Sevilla, 315 pp.

Leira, E. and Quero, D. (1988) *Calvià a Bon Port. Avance del Plan de Ordenación del Municipio de Calvià*. Ajuntament de Calvià, Mallorca, 176 pp.

Ley 2/89, de la Generalitat Valenciana, sobre Ordenación del Territorio.

Marchena, M. (1988) La estrategia territorial de la nueva política turística de Andalucía. *Urbanismo* 4, 55–65.

Marchena, M. (1990) Implicaciones territoriales de la política turística en Andalucía. In: Cano García, G. (ed.) *Geografía de Andalucía*. Tartessos, Sevilla, pp. 328–348.

Pedreño Muñoz, A. (ed.) (1986) *Libro Blanco del Turismo en la Costa Blanca*, 2 vols. Cámara Oficial de Comercio, Industria y Navegación de Alicante, Alicante.

Quero, D. and Leira, E. (1990) *Avance del Plan Insular de Ordenación de Gran Canaria*. Excmo. Cabildo Insular de Gran Canaria, Las Palmas de Gran Canaria, 159 pp.

Secretaría General de Turismo (1992) *Plan Marco de Competitividad de Turismo Español*. Ministerio de Industria, Comercio y Turismo, Madrid, 125 pp.

Torres Bernier, E. (1989) *Libro Blanco del Turismo de la Costa del Sol Occidental*, 12 vols. Dirección General de Turismo, Sevilla.

Urry, J. (1990) *The Tourist Gaze: Leisure and Travel in Contemporary Societies*. Sage Publications, London, 176 pp.

Vera, F. and Monfort, V. (1994) Agotamiento de modelos turísticos clásicos. Una estrategia territorial para la cualificación: la experiencia de la Comunidad Valenciana. *Estudios Turísticos* 123, 17–46.

Vera Rebollo, F. (1990) Turismo y territorio en el litoral mediterráneo español. *Estudios Territoriales* 32, 81–110.

Vera Rebollo, F. (1992) El modelo turístico: características y cambio. In: Velarde, J., García Delgado, J. and Pedreño, A. (eds) *VI Jornadas de Alicante sobre Economia Española, Ejes Territoriales de Desarrollo: España en la Europa de los Noventa*. Colegio de Economistas de Madrid, Madrid, pp. 421–443.

Vera Rebollo, F. and Marchena, M. (1990) Turismo y desarrollo: un planteamiento actual. *Papers de Turisme* 3, 59–84.

9 Tourism and Environmental Degradation: *the Northern Adriatic Sea*

GABRIELE ZANETTO AND STEFANO SORIANI

Economic Development and the Environment in the Adriatic Coastal Region

The Adriatic Sea, in particular, the northern Adriatic, constitutes one of the most problematic areas of the Mediterranean Sea as far as the relationship between economic development and environmental issues is concerned. This is largely the result of a combination of three factors: the uniqueness of the environmental conditions in the basin, the large quantity of pollutants discharged into it and the steadily increasing social demands placed upon this basin and its resources. As a result, environmental quality has become a crucial issue in local social and political life.

Physical and environmental features

The Adriatic Sea, with a surface area of approximately 136,000 km^2 and a volume of approximately 35,000 km^3, is traditionally divided into three sub-basins: the northern Adriatic, north of a line from Ancona to Zara, comprising 20.3% of the total; the middle Adriatic, whose southern boundary is the line from Gargano to Neretva – 37%; and the southern Adriatic, which extends southwards to the Straits of Otranto, making up the remaining 42.7% (Fig. 9.1). The characteristics of the northern basin are highly specific. While it accounts for about one-fifth of the total surface area of the Adriatic Sea, it accounts for only 3% of its total volume – mean depth, 30 m. More than three-quarters of the fresh waters from the rivers in the catchment basin flow into this sub-basin. The northern and the middle Adriatic sub-basins account for one-third of the Mediterranean's entire supply of fresh water, which flows into the basin from a network of rivers extending from

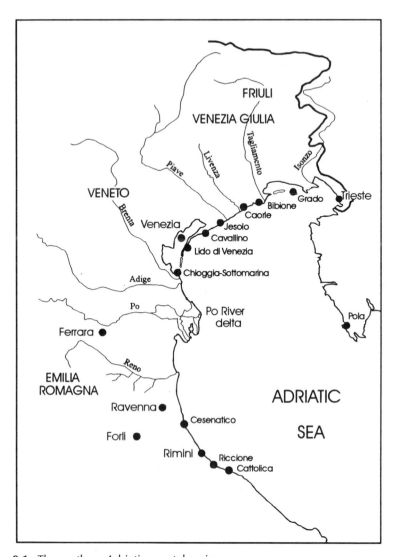

Fig. 9.1. The northern Adriatic coastal region.

Isonzo in Friuli to Fortore in Puglia. About 50% of these fresh waters come from the River Po; 15% from the rivers that flow south of it; and a further one-third from the rivers that flow to its north. After prolonged rainfall, the inflow of water to the basin may triple. This remarkably abundant inflow, mainly adjacent to the Po delta, explains not only the low level of salinity in the Adriatic in comparison with other Mediterranean basins, but also the abundance of fish present, fed by the many nutrients washed down by the rivers (Bombace, 1985; Marchetti, 1994).

The hydrodynamics of the Adriatic coastal region are particularly complex and varied, depending on whether the coastal or the open-sea subsystems, northern, middle or southern sub-basins, the winter–spring or summer–autumn seasons are taken into account. Consequently, any fore-cast of the potential action of substances flowing into the sea is fraught with uncertainty (Franco, 1985). The hydrodynamism of the northern basin is unique in that, mainly during the summer months, coastal waters tend to become segregated from open-sea waters, thus favouring stagnation pro-cesses in pollutants along the coast. Moreover, the vertical layering of waters is a common occurrence in summer and, under particular climatic con-ditions – for instance, the absence of east or northeast winds. This can give rise to anoxia and eutrophication (Marchetti, 1994). The extension of land surrounding the sea, comparable to that of other semiclosed seas in Europe, its shallow mean depth (40 m from Rimini to Pola), the richness of nutrients, the high biological productivity and the wide variety of the natural environ-ments make the Adriatic basin one of the most fragile in the Mediterranean (Vallega, 1989).

Economic activities in the Adriatic Sea

The Adriatic constitutes an important resource for a number of economic activities, tourism being but one of them. The interdependence of these activities reflects both the competition for coastal areas and the effects that these activities have on coastal environments.

The northern Adriatic is the most important area of offshore extraction of natural gas in Italy. This activity began in 1959, and since then more than 500 wells have been drilled, half of which have entered the production phase. Whilst prospection was at first limited to the marine areas closest to the shore of Ravenna, it was subsequently extended to include the deepest areas of the open sea. Today, offshore activities have been pushed to the shelf limit defined by international agreements, based on the median line principle. Development programmes foresee an increase in activity in the more populated waters of the Ravenna area and the extension of drilling activities towards the southeast, around Ancona, and towards the northwest in the area around Chioggia (Soriani and Zanetto, 1992). As for the environ-mental impacts of prospection, no problems of pollution have been recorded. The most worrying environmental problem is the subsidence of the seabed where methane gas has been extracted. In some marine zones adjacent to Ravenna, for example, the seabed has subsided 130 cm over a period of 30 years. At Lido Adriano, near Ravenna, the lowering of the seabed by about 50 cm has caused the shoreline to recede 120 m (Carbognin, 1986). Cur-rently, proposed projects for the extraction of methane gas in marine areas adjacent to the Venice lagoon are causing concern.

The Adriatic Sea, Italy's most productive fishing basin, is the origin of more than half the national fish catch. The northern sub-basin is the most productive area (Bombace, 1985), given its shallow depth – its maximum depth is a little over 70 m – and its richness in nutrients. Many new projects in the field of aquaculture are in the process of being implemented. The fishing industry, however, is suffering the consequences of eutrophication, pollution and changes in the biotic and abiotic characteristics of many breeding zones, as well as overfishing and drag-netting by the fishing industry itself (Fondazione Giovanni Agnelli, 1990).

The Adriatic constitutes an important transit corridor for navigation. More than 30 harbours are located along the 1150-km-long Italian Adriatic coastline, which currently handles a maritime traffic flow of approximately 80 million tons per year. Oil products constitute an important part of this traffic – over 40% of the total traffic in the port of Venice, more than 30% of the traffic in Ravenna and in excess of 80% of the traffic in Trieste (ICHCA, 1991). Despite the increasing importance of break bulk, liquid and dry bulks still remain the main traffic segments. The problems of the dangers of navigation through this corridor and pollution caused by maritime traffic should be considered within the context of the general evolution of the Adriatic city-ports. It should be remembered that the transition of many Adriatic port and industrial areas to Maritime Industrial Development Areas (MIDAs) of the fourth generation, with advanced-technology industrial activities and services and less impact on the environment, has not yet been completed (Vallega, 1992). Finally, it should be noted that, until a few years ago, no control was exercised over the amount of silt dredged from port basins and channels, which was systematically discharged into the Adriatic (Fondazione Giovanni Agnelli, 1990).

The Adriatic is one of the basins which contributes to telluric pollution in the Mediterranean (Fondazione Giovanni Agnelli, 1990). This form of pollution, which adds to the complexity of the environmental conditions in the area, is caused by the large inflow of fresh water to the basin from the catchment area and the intensity of economic activity. The surface area of the catchment basin of the Adriatic is approximately 220,000 km^2, with a population of approximately 40 million inhabitants. About 70% – 155,000 km^2 with a population of approximately 34 million inhabitants – lies in Italian territory, and comprises half the surface area of Italy. Pollution caused by human activities in the area is equivalent to that normally generated by about 200 million inhabitants: 43.8% of this pollution originates from industrial activities, 16.7% from civil uses, 29.6% from agriculture and 9.9% from animal husbandry. Most of this organic pollution – approximately 70% – originates in the Po basin (Ghetti, 1992), which accounts for 26% of the total surface of Italy and about 32% of the national population – 18.4 million inhabitants. This area generates 42% of Italy's gross national product (GNP), about 32% of the nation's income from agriculture and more than 50% of its income from industry.

Tourism Development in the Adriatic Area: Main Features and Recent Trends

The 'seaside holiday' in the northern Adriatic – which became popular during the 1950s and 1960s as more free time and higher real incomes became available – progressively replaced the élite nature of tourism existing up to that time. Three main factors determined its entry into the mass tourism market: its proximity to areas where the demand for tourism was growing, generally the areas of greatest economic development – Germany, Austria, Switzerland and northwestern Italy; transport infrastructure providing easy access to the area; and the existence of natural resources to meet the needs of tourist demand, which was essentially limited to 'sun, sand and sea' motivations (Vecchio, 1988; Gambuzza and Sartore, 1993).

General characteristics of seaside resorts

With the exception of the Po River delta, the upper-Adriatic coastline from Grado to Rimini has been subject to substantial tourism development. Broadly speaking, two different types of tourism-generated coastal environments exist in the area, each of which reflects the different periods in which tourism industry development occurred as well as the composition of demand from both Italian and foreign tourists (Bernardi *et al.*, 1989; Politi and Preger, 1991).

The first type is found mainly on the Romagna coast, south of Ravenna. Here tourism development progressively occupied the areas closest to the sea. Urbanization took place with no concern for continuity. Tourism in the area in the interwar period and even earlier was largely élitist and its development was chiefly hotel-based. Italian tourists constituted its most important component. The second type is found along the Friuli–Veneto–Ferrara coastline, stretching northwards from Ravenna to Grado, where tourism has developed more recently. The physical environment of the area – lagoons, marshes and estuaries – has impeded the continuity of resorts, where private residences – rooms or flats for rent, holiday homes, etc. – and, more recently, camping sites have become the commonest forms of tourist accommodation. Foreign tourists are the main component of demand, and Germans constitute the largest group.

Two important resorts do not come within these two categories: Jesolo and the Venice Lido. The importance of hotel accommodation in Jesolo and the fact that the tourist industry developed earlier here are the two factors which make it similar to the resorts on the Romagna coast. Venice Lido, on

Table 9.1. Arrivals and nights spent in the Province of Forlì, 1985–1993 (absolute figures in thousands, rounded to nearest thousand) (from ISTAT, 1987–1995).

	1985	1986	1987	1988	1989	1990	1991	1992	1993
Arrivals–Italians	1,641	1,770	1,885	1,931	1,704	1,920	2,130	2,174	2,245
Arrivals–Foreigners	692	765	804	765	537	473	567	521	518
Arrivals–total	2,333	2,535	2,689	2,696	2,241	2,393	2,697	2,695	2,763
Change in %	–	+8.6	+6.1	+0.3	-16.9	+6.8	+12.7	-0.1	+2.6
Nights–Italians	15,283	15,678	15,823	16,076	11,525	12,836	14,772	14,763	14,608
Nights–Foreigners	6,486	7,008	7,157	6,814	4,311	3,380	4,192	3,689	3,576
Nights–total	21,769	22,686	22,980	22,890	15,836	16,216	18,964	18,452	18,184
Change in %	–	+4.2	+1.3	-0.4	-30.8	+2.4	+16.9	-2.7	-1.5

Table 9.2. Arrivals and nights spent in the Province of Venice, 1985–1993 (absolute figures in thousands, rounded to nearest thousand) (from ISTAT, 1987–1995).

	1985	1986	1987	1988	1989	1990	1991	1992	1993
Arrivals–Italians	1,383	1,427	1,381	1,511	1,416	1,358	1,389	1,392	1,388
Arrivals–Foreigners	2,720	2,514	2,789	2,778	2,517	2,586	2,399	2,533	2,721
Arrivals–total	4,103	3,941	4,170	4,289	3,933	3,944	3,788	3,925	4,109
Change in %	–	-4.0	+5.8	+2.9	-8.3	+0.3	-4.0	+3.6	+4.7
Nights–Italians	9,961	9,864	9,803	9,885	8,290	6,196	6,918	7,202	6,950
Nights–Foreigners	13,086	13,485	14,602	14,549	11,263	9,148	9,805	10,121	11,042
Nights–total	23,047	23,349	24,405	24,434	19,553	15,344	16,723	17,323	17,992
Change in %	–	+1.3	+4.5	+0.1	-20.0	-21.5	+9.0	+3.6	+3.9

the other hand, is a tourist resort that is unique to the upper Adriatic. This is because the Lido is a residential neighbourhood of Venice's historical centre and, as a result, its evolution reflects various aspects of Venice's urban development which are unrelated to tourism. Moreover, Venice Lido has always represented a unique kind of resort which satisfies tourists' desire for both a 'seaside' and a 'cultural' holiday.

It is not easy to provide a statistical summary of the importance of tourism in the upper Adriatic. Tables 9.1 and 9.2 show visitor arrival and bed-night statistics for the period from 1985 to 1993 in the provinces of Forlì and Venice. These demonstrate well the characteristics of the two types of tourist development mentioned above. Forlì, Cesenatico, Rimini, Riccione and Cattolica are the most important resort communities – although it should be noted that Rimini is now a province in its own right. Italians account for the largest proportion of tourists in these areas – 80.3% in 1993. A more complex pattern of tourism development exists in the province of Venice, due to the existence of the city's historic centre, which attracts cultural tourism while also acting as a magnet for nearby seaside resorts (Gambuzza and Sartore, 1993). The most important of these are Chioggia-Sottomarina, Venice Lido, Cavallino, Jesolo, Caorle and Bibione. In 1993, 61% of tourists in the province were from outside Italy, an overall figure which, however, belies important variations amongst the resorts: in Jesolo and Caorle foreign tourists constitute an average of about 50% of the total tourist influx; in Chioggia-Sottomarina the average figure is approximately 20%; at Venice Lido the figure tops 60%; Bibione and Cavallino are largely dominated by cross-border tourism, which accounts for, on average, 60 and 65–70% of visitors respectively.

Resort maturity and restructuring of tourism activities

Today the seaside resorts of the upper Adriatic have to contend with new problems deriving mainly from the effects of two general tendencies (Guidicini and Savelli, 1988; Van der Borg and Gambuzza, 1992):

- changing trends in tourist behaviour, characterized by a reduction in the average length of holidays and greater attention to the quality of natural resources, services offered and value for money;
- the emergence on the market of new seaside holiday resorts, as a result of regional policies for areas which were economically disadvantaged, but which had relatively intact natural environments to offer; lower prices at resorts which were not yet well established in the market; and the impact of large tour operators and air transport on tourist flow patterns.

Following a period of continuous growth, which lasted until the second half of the 1980s, the ability of the upper-Adriatic resort communities to compete in the tourist market has consistently been undermined in a market which is becoming increasingly more complex. A restructuring process is now under way in the major upper-Adriatic resorts, which aims to combat three closely interrelated problems: environmental degradation and overcrowding; the need to improve the quality of existing services; and the need to diversify services and resources offered.

The following specific objectives have been established:

- improvement in overall quality and value for money of the services offered;
- closer cooperation with large tour operators – clearly only in resorts where the tourism industry is well developed and a less fragmented accommodation supply makes it possible to pursue this objective;
- lengthening of the tourist season through dynamic price policies;
- research to pinpoint new sectors of demand for the low season;
- improvement in the quality of life at the tourist resorts by reducing overcrowding and traffic congestion, providing adequate parking, improving the quality of tourist accommodation and providing access to 'green' areas;
- diversification of local tourist services and promotion of a new attitude to tourists that is less dependent on the traditional image of holidays for 'rest and relaxation'.

The aim of providing an attractive natural environment is another important factor in this restructuring process. For example, some seaside resorts of Veneto offer nature-orientated activities as an attraction, making use of the coastal topography for cycle paths and nature walks, creating small parks and promoting guided visits to some lagoons in the interior, etc. The seaside resorts on the Romagna beaches are being promoted in conjunction with the added attraction of the nearby artistic/cultural cities in the valleys and hills of the Apennines.

This already complex picture was further complicated in the second half of the 1980s by the issue of eutrophication. Indeed, the widespread appearance of phenomena of mucilage (*mucillagine*), associated with 'cloudy sea' or 'marine snow', in many parts of the upper-Adriatic coastal waters in the summers of 1988 and 1989 was probably brought about by unicellular algae (*Diatoma, Cyanophyceae*), whose existence can be traced back to the 1800s (Fondazione Giovanni Agnelli, 1990; Marchetti, 1994). Whatever the scientific explanation, this occurrence, coupled with alarmist information spread by the mass media, caused a drop in the number of tourists to the area of the order of 20–30% (Tables 9.1 and 9.2).

The drop in the number of arrivals recorded in 1989 has since been reversed, but this is not true for the figures of overnight stays, which have suffered an overall reduction in number, thus confirming the trend towards shorter average holiday length. However, upper-Adriatic seaside tourism currently appears to be showing renewed vigour, which is the result of a new set of factors.

- 1989 was the most critical year for tourism, largely as a result of the appearance of mucilage, but this phenomenon has not recurred since to the same extent. Indeed, the condition of coastal waters with regard to eutrophication seems to have improved in the past few years, due to the implementation of various policies – water treatment, reduction of phosphorus content in detergents, etc.
- Since 1989 the quality of information available on the condition of the upper-Adriatic environment has greatly improved and this has had a notable effect on tourist behaviour.
- Eutrophication and mucilage worsened a situation that was already becoming critical, since many northern Adriatic resorts were now entering the maturity stage. The most dynamic seaside resorts attempted to react by undertaking various measures to improve their services, by advertising and by adopting flexible price policies. Within the framework of current restructuring efforts many resorts have sought to reduce their dependence on the natural condition of the coastal waters by installing mobile barriers to keep algae away from the beaches and investing in complementary services – swimming-pools on the beaches, water-fun-based theme parks, etc.

The role of individual factors within the overall scenario cannot be fully appraised at the present time. It is clear that the positive results of the past few years also depend on uncontrollable local and regional factors, which include the continuous devaluation of the Italian lira in relation to other currencies and the effects of international events, such as the Gulf War and the outbreak of hostilities in the former Yugoslavia, and international tourism trends in the Mediterranean.

Tourism Development and Coastal Environments in the Adriatic Area

With the exception of the Po delta, where agriculture, fishing and fish farming have always been the prevalent activities, tourism has brought about profound changes in the coastal environment of the upper Adriatic, as a result of a constantly growing influx of tourists. The nature of the impact of tourism has mostly depended on certain characteristics of the demand during the years of greatest increase in the number of tourists, when seaside holidays for rest and relaxation were in vogue. The main demand was thus for easy access to safe beaches and the sea. Gradually shelving beaches and the presence of a broad sandy shore, often created by levelling coastal dunes, easily accommodated this demand. This led to an intense and uncontrolled process of coastal urbanization during the 1950s and 1960s, assisted by the absence of an efficient coastal planning system and a lack of public concern about construction quality and environmental protection. The most evident result of this 'race to the coast' was the disappearance of the area's characteristic coastal dunes and pine copses. The beaches have become highly artificial environments, which have to be maintained by carrying out continual work. This is a task that is proving increasingly difficult as erosion takes its course, and which has required progressively more extensive and complex coastal defences (Fabbri, 1989).

Nevertheless, tourism is just one of the industries which has contributed to the changes in coastal regions during this century. In many cases the processes of urbanization, the expansion of agriculture and the development of ports and industrial areas connected with basic industries greatly altered the coastal environments long before tourism assumed its role as the driving force behind the spatial organization of coastal areas. The most conspicuous example can be seen in the beaches of Veneto. Here the coast has always been an extremely ill-defined entity. Hydraulic interventions carried out over many centuries established the morphological equilibrium of the Venice lagoon and gave form to the coastline. Reclamation operations continued well into this century, for reasons of health and hygiene and to aid agricultural development, and were finally concluded in the interwar period. The construction of roads on land reclaimed from coastal marshes led to further urbanization processes along the coast and the appearance of the first forms of social tourism for young people, through the introduction of summer camps. These initiatives led to the mushrooming of new tourism settlements in various parts of the area shortly after the Second World War. Later, in the 1960s, tourism was further developed, but it maintained strong ties with agricultural activities – tourism became a source of supplementary income for many farmers during the summer months, as it could be exploited with very little capital investment. Tourism development in the 1960s and 1970s took place in coastal regions which had already been greatly influenced by humans, from both an environmental and a morphological viewpoint, and only then did it become the main protagonist in the spatial and functional organization of many areas.

Notwithstanding the absence of an adequate monitoring system, tourism is considered an activity that has contributed to the deterioration in the quality of Adriatic coastal waters, mostly due to its highly seasonal nature (Fondazione Giovanni Agnelli, 1990). A comparison between the permanent resident population and the peak of the summer months allows a first assessment of the problem to be made: Cesenatico has about 20,000 inhabitants; Rimini 130,000; Riccione 33,000; Cattolica 15,000; Jesolo 22,000; Caorle 11,000; Bibione has less than 10,000; Cavallino about 11,000; the Venice Lido 19,000; Chioggia-Sottomarina about 53,000; and the historic centre of Venice has about 70,000 residents, together giving an aggregate total of

394,000 people. Yet these places register an annual average of more than 35,000,000 tourist bed-nights.

Water-related environmental impacts

The impact of this burden obviously depends on many factors, the most important being hydrodynamics and the pattern of marine currents, the condition and efficiency of sewage systems and water treatment systems, prevailing legislation and the efficacy of its implementation. With regard to the first aspect, the pattern of currents in the upper Adriatic – especially during the summer – creates a situation whereby the coastal waters remain relatively segregated from those further offshore, thus favouring stagnation processes in organic waste substances. The water adjacent to the beaches often presents critical situations. There, pollutants are most heavily concentrated, especially near the highly populated areas or river mouths, and water circulation is reduced even further as a result of the presence of coastal defences, constructed to make the area more accessible and to counteract erosion.

Information regarding sewage systems and water treatment in coastal communities, especially the smaller ones, is unreliable. It can, however, be asserted that the coastal communities of the upper Adriatic shared the same fate as other Italian communities up until the mid-1980s. An inquiry carried out at the time showed that, in the three regions of the upper Adriatic – Friuli–Venezia–Giulia, Veneto and Emilia Romagna – the town councils equipped with waste treatment plants amounted to 70, 59 and 73% respectively. None the less, according to another inquiry conducted at national level, only about 35% of waste purification plants functioned properly (Ghetti, 1992). Although the first coordinated Act on water and waste regulation was passed in 1976, for many years administrative obstacles and poor organization within the departments appointed to implement it impeded the effective application of the law.

Thus a reliable estimate of the impact of tourism on water quality cannot be made, as this would require a much more precise level of analysis. The above studies do, however, show that up until at least the middle of the 1980s very little attention had been given to waste management, the condition of coastal waters or even the effects tourism had on the quality of coastal waters. Confirmation is provided by Italy's tardy acknowledgement of the European Economic Community (EEC) directive 76/160 (EC, 1976) on water conditions at seaside resorts only in 1982. The first water samples to be examined were not taken until 1984 and, as late as 1987, more than half of the Italian coastal waters had still not been adequately analysed (Fondazione Giovanni Agnelli, 1990). The situation changed towards the end of the 1980s. In particular, the controversy arising from 'the algae summer' of 1989 brought the Adriatic and the condition of its coastal waters into the national spotlight. New research programmes showed that coastal waters were often in a critical condition due to their high content of organic pollution, nutrients and, close to the river mouths, heavy metals and hydro-carbons. At the same time, although certain important parameters, such as eutrophication, were not acknowledged by the Italian system, a more strin-gent application of the EEC directive 76/160 led to a temporary bathing ban at some of the most popular beaches of the upper Adriatic.

The tourism and fishing crisis dramatically illustrated the strategic role of water quality in an extremely complex coastal region. The experience acquired over the years showed how the increase in the demands society placed on the sea, both directly as a supplier of natural resources and indirectly as a waste receptacle, had transformed the environmental quality of the coastal waters into strategic bargaining counters in the confrontation between various social factions. In this context, it could be said that tourism has increasingly played the role of an 'ambassador of protection' for the environment of the upper Adriatic. In many respects, the tourism sector has proved to be an ally of environmentalist movements and has provided sup-port for their most important initiatives. The case of eutrophication offers ample evidence of the role tourism can play in promoting marine environ-ment protection in a coastal region subjected to various forms of economic use that are potentially incompatible on account of their environmental impacts.

Tourism and environmental conflict: the case of eutrophication

Eutrophication is a very frequent phenomenon in the northern Adriatic sea. As mentioned earlier, this depends on its specific environmental conditions – shallow depth, the great quantity of fresh water poured into it by the catch-ment basin and its unique hydrodynamism. The problem of eutrophication in Italian seas, however, began to gain importance from the mid-1970s onwards. Eutrophic processes have since intensified and persisted over longer time spans, with a tendency towards progressive expansion (Marchetti, 1994). The worsening situation was largely brought about by the increase in nutrients in the coastal waters. On a national scale, coastal and inland waters receive about 48,000 tons year^{-1} of phosphorus and 660,000 tons year^{-1} of nitrogen. It has been estimated that the coastal waters of the north-ern Adriatic receive just under 50% of the total amount of phosphorus deposited in Italian waters (Ghetti, 1992). The most critical situation is en-countered south of the Po River delta near Ravenna in a marine area extending for about 100 km, where the polluting effects of the Po's waters during the summer season are particularly serious (Marchetti and Rinaldi, 1989). Considering the total expanse of the Adriatic Sea, this is, in fact, a relatively small area in absolute terms. The fears expressed in 1989 by environmentalists, who spoke of the death of the Adriatic, have thus proved unfounded. The general state of deterioration of the Adriatic waters has, indeed, improved since the end of the 1980s. This is true with regard not only to oligotrophy – lack of nutrients – in open sea waters, but also to the prevailing condition of coastal waters.

Eutrophication seemed destined to deal an almost mortal blow to the tourist expectations of the Adriatic, the most important area for seaside tourism in Italy, characterized by the stage of maturity of the sun/sand/sea market and subject to competition from new destinations. But this has not, in fact, happened. The importance of tourism in the Adriatic area has therefore contributed to the development of more responsible behaviour and attitudes, largely as a result of the protests of the different social groups involved in tourism activities in the late 1980s. These have led to the reduction of phosphorus content in household detergents; the provision of efficient waste water treatment plants in coastal cities; the introduction of new eutrophy monitoring programmes; the setting up of new bodies to inform the public of environmental conditions in the Adriatic; the suspension of large-scale dumping operations in northern Adriatic areas; and discussions on the eventual 'recovery' of the River Po.

However, the problem of the relationship between tourism and the protection of coastal environments is complex, not only because tourism itself produces important environmental impacts, but also because only some components of a basin's ecosystem contribute to define the environmental quality on which the tourists' evaluation is based. Moreover, tourism, like other economic activities, tends to impose its own order on the coastal environment, moulding it to the prevailing tourism model. Cultural and functional elements combine to define what is worthy of protection. Although this statement is generally true, in the Adriatic context the relationships between tourism and other direct and indirect uses of the sea are characterized by a latent conflict. Moreover, it must be remembered that the EEC directive 76/160 on the quality of coastal water required for bathing has been loosely interpreted by the Italian government (Fondazione Giovanni Agnelli, 1990). Stricter regulations concerning the nature of regional differences in interpretation – dealing with aspects such as water transparency or eutrophic parameters – would make the existence of the conflict more obvious.

Clearly, to adopt a functional economic criterion to define the economic and social value of the environment only in terms of social demand is eminently reductionist. Its justification lies in: the absence of the information necessary to address the problem of assessing the state of ecosystems; excessive consideration of present preferences in planning for the future; the short-sightedness of policies based on economic functionality that undervalue long-term environmental damage, etc. Thus, despite the fact that there is evidence that further damage could create a biologically very different, impoverished basin, it is not difficult to understand why none of the economic activities – including fishing – has wished to raise the voice of alarm for the protection of 'Posidonia's meadow'. On the other hand, social demand has brought about steadily increasing sensitivity to environmental conditions. This is the context which serves as the basis of strategies in the process of coastal maritime regionalization in the northern Adriatic (Zanetto, 1992), and has centred the social debate on the problem of the radical changes that each activity tends to impose upon marine environments.

Conclusions

The chaotic and poorly planned development of tourism during the past few decades has brought about a drastic reduction in the quality of the environmental resources on which it must inevitably depend, but the possibility of sustaining seaside tourism at its present level of development largely depends on the quality of these resources. At the root of this question lies the changing pattern of motivations in tourism. Such changes in market trends make it necessary to place greater emphasis on environmental issues: a harmonious environment and higher quality of life are the most essential elements in attracting seaside tourism without further impoverishing local resources. In this context, local politicians in seaside resorts are now paying a great deal of attention to environmental issues. Urbanization has currently reached peak levels and new pro- grammes to improve the quality of tourist accommodation are being implemented. Drinking-water is monitored with greater care; differentiated solid waste collection is being introduced in all of the cities; the creation of 'green' urban spaces is a feature common to all of the resort communi- ties. Indeed, some resorts are even investing in 'nature': areas considered marginal – such as small, marshy coastal zones – have become the object of initiatives to create parks hailed as 'natural oases', within the context of marketing strategies aimed at promoting a new tourist image, attracting new segments of tourist demand and reducing the seasonal nature of the tourism industry. This is clearly a functional approach to the problems of the coastal environment: the environment is considered to be one of the fundamental features which define the current restructuring process taking place within the tourism industry and, at the same time, one of the most important elements in the marketing of resorts.

   It is more difficult to appraise the role of tourism as a source of environ- mental damage. Seaside tourism in the upper Adriatic has undoubtedly con- tributed to altering the features of the coastal environment, but this has occurred within a very complex territorial setting, in which other factors – namely agriculture, ports and coastal industries, and urbanization in gen- eral – are also involved. Bearing in mind the elements noted earlier, it is obvious that the northern Adriatic Sea is expected to perform too many functions at the same time: it is an important area for recreation and leisure; it is one of the most important areas in Italy for the direct or indirect disposal of waste; it is one of the most important reservoirs of biological resources in the Mediterranean; it is a very important area for aquaculture; it is a transit corridor for constantly growing navigation flows on which projects for greater integration of the economy and its markets in the future are based; it is a suitable area for the exploitation of underground resources; and it is still a strategic space from a geopolitical point of view.

In these circumstances, it has proved extremely difficult in recent years to ensure an environmental quality which is consistent with the different social demands placed upon the sea and with their different sensitivities to the condition of marine environments. The need for maintaining the coexistence of these activities, which is the problem of ensuring the sustainability of the Adriatic coastal region considered as a complex system, means that adequate coastal management approaches must be adopted. No single use of the sea can be considered more important than the overall environmental quality of the Adriatic basin. Moreover, considering that the condition of coastal ecosystems is, at the same time, the final result of the activities carried out upon adjoining inland areas, it is clear how extensive the horizon of the problems to be faced actually is. Indeed, the recent difficulties in debating the subject of agricultural pollution in the Po basin and the problems faced by many small firms in central and northeast Italy in adopting methods capable of reducing water pollution are proof of that. However, in the absence of an integrated approach, the Adriatic is likely to become a poorer basin in the future and less and less capable of meeting the social demands placed on it. In this context, the development of seaside tourism has served to bring once more to the fore the issue of the relationship between the land and coastal environments in political and social debate. The reconsideration of the environment on the grounds of strictly functional economic criteria shows important ambiguities but, notwithstanding, it is on this basis that some positive results have been achieved recently in the northern Adriatic coastal region.

References

Bernardi, R., Lando, F., Marinucci, M. and Zanetto, G. (1989) L'alto Adriatico: articolazione di uno spazio costiero. In: Bernardi, R. (ed.) *Mari e Coste Italiane*. Pàtron, Bologna, pp. 153–188.

Bombace, G. (1985) Eutrofizzazione e produttività ittica dell'Adriatico. In: De Carlo, G. (ed.) *Lo Smaltimento dei Rifiuti Industriali ed i Problemi dell'Alto Adriatico*. Gruppo Montedison, Milano, pp. 70–77.

Carbognin, L. (1986) La subsidenza indotta dall'uomo nel mondo: i casi più significativi. *Bollettino dell'Associazione Mineraria Subalpina* 4, 433–468.

EC (European Commission) (1976) EU Bathing Water Directive, 76/160/EEC. *Official Journal of the European Communities*, EEC, Brussels.

Fabbri, P. (1989) Lo spazio-spiaggia: usi ed erosioni. In: Pranzini, E. (ed.) *La Gestione delle Aree Costiere*. Edizioni delle Autonomie, Roma, pp. 120–135.

Fondazione Gioavanni Agnelli (ed.) (1990) *Manuale per la Difesa del Mare e della Costa*. Edizioni della Fondazione Giovanni Agnelli, Torino, 289 pp.

Franco, P. (1985) Caratteri strutturali e di circolazione dell'Adriatico settentrionale: relazioni con la diffusione di soluti versati nel bacino. In: De Carlo, G. (ed.) *Lo Smaltimento dei Rifiuti Industriali ed i Problemi dell'Alto Adriatico*. Gruppo Montedison, Milano, pp. 22–27.

Gambuzza, M. and Sartore, M. (eds) (1993) *Forme e Processi di Valorizzazione Turistica*. Franco Angeli, Milano, 299 pp.

Ghetti, P.F. (1992) *Manuale per la Difesa dei Fiumi*. Edizioni della Fondazione Giovanni Agnelli, Torino, 293 pp.

Guidicini, P. and Savelli, A. (eds) (1988) *Il Turismo in una Società che Cambia*. Franco Angeli, Milano, 254 pp.

ICHCA (International Cargo Handling Coordination Association) (ed.) (1991) *I Porti Italiani*. Trasporti Industriali e Movimentazione ETAS, Milano, 78 pp.

ISTAT (Istituto Centrale di Statistica) (1987–1995) *Statistiche del Turismo*, Istat, Roma, annual publication.

Marchetti, R. (1994) *L'Eutrofizzazione. Un Processo Degenerativo delle Acque*, 3rd edn. Franco Angeli, Milano, 315 pp.

Marchetti, R. and Rinaldi, A. (1989) Le condizioni del mare Adriatico. In: Melandri, G. (ed.) *Ambiente Italia 1989. Rapporto della Lega per l'Ambiente*. ISEDI, Torino, pp. 33–37.

Politi, U. and Preger, E. (1991) Modelli di sviluppo turistico. In: Fuà, G. (ed.) *Orientamenti per la Politica del Territorio*. Il Mulino, Bologna, pp. 381–411.

Soriani, S. and Zanetto, G. (1992) *Northern Adriatic and Offshore Activities: Features and Onshore Impacts*, Nota di lavoro 11.92. Fondazione ENI Enrico Mattei, Milano, 14 pp.

Vallega, A. (1989) Mari italiani e mediterraneo. In: Bernardi, R. (ed.) *Mari e Coste Italiane*. Pàtron, Bologna, pp. 9–25.

Vallega, A. (1992) *The Changing Waterfront in Coastal Area Management*. Franco Angeli, Milano, 128 pp.

Van der Borg, J. and Gambuzza, M. (1992) Le spiagge venete: 'winners' o 'losers' nell'attuale quadro del turismo balneare mondiale? *Coses Informazioni* 1, 13–20.

Vecchio, B. (1988) Valorizzazione ed innovazione territoriale: riflessioni sul caso delle aree turistiche italiane. In: Leone, U. (ed.) *Valorizzazione e Sviluppo Territoriale in Italia*. Franco Angeli, Milano, pp. 157–170.

Zanetto, G. (1992) La régionalisation marittimo-littorale: un cas de structuration dans l'Adriatic du Nord. In: Lando, F. (ed.) *Urban and Rural Geography. Papers from 6th Italian–Polish Geographical Seminar*. Cafoscarina, Venezia, pp. 189–195.

10 Tourism and Carrying Capacity in Coastal Areas: *Mykonos, Greece*

HARRY COCCOSSIS AND APOSTOLOS PARPAIRIS

Introduction

In recent years there has been considerable concern about the efficacy of the available planning and management tools in controlling or regulating human actions in such sensitive environments as coastal zones. This chapter examines the key development and environmental problems existing in coastal areas and the causes leading to conflict, and recognizes the importance of both human and physical environmental systems, while advocating a new way of balancing their conflicts within the framework of sustainable development, using the case-study of Mykonos. The concept of carrying capacity is examined, as a methodological tool to protect resources and respect interest constraints in the course of supporting economic development, including tourism, and to protect the natural and human-made environment, especially in coastal areas.

Since the Second World War one of the most rapidly developing sectors of the world economy is that of tourism, a phenomenon which has taken on a mass character and has expanded to such a scale that, in many cases, the negative effects already exceed acceptable levels of disturbance in sensitive environments, such as coastal areas (Coccossis and Parpairis, 1995). Various studies carried out recently noted major changes in the coastal system as a result of the considerable pressure of human activity and particularly of tourism, a sector which is regarded as having important negative effects on the coastal environment mainly because of its mass and seasonal nature and the pattern of its geographical distribution (Miossec, 1977; Pearce, 1989; Nijkamp *et al.*, 1991; DoE, 1993; Coccossis and Parpairis, 1995).

It is possible to observe changes in demand, thus obliging tourism to enter a new phase of development, in which the main objective is the diversification of the product from mass and package tourism towards new forms, including active, selective, culture-orientated and soft and ecotourism, which use as their banner the quality of services and the protection of the environment. As a consequence of the mass tourism development of the past 30–40 years, cities, towns and villages located in coastal areas have faced considerable challenges and pressures for development, which have altered their sensitive coastal and heritage features, so attractive to residents, tourists and business visitors alike, including rural setting, coastal development, seaside development at port facilities, etc. Such pressures have been common in Europe, especially in the Mediterranean coastal and island regions, and obviously in the Greek setting with its numerous islands and extensive coastal areas.

However, as has already been pointed out, it is well known that one of the main problems related to tourism in coastal ecosystems is the effectiveness of existing administrative, legal and management frameworks that have evolved to control or regulate activities in the coastal zone. These concerns have been articulated by many interest groups – by local, national and international governments as well as other responsible bodies, such as marine conservation societies and wildlife and environmental organizations. In this respect, coastal planning and management serves as a framework within which to attempt to present evidence of the nature of existing coastal problems, and to provide views about the efficacy of the planning and management tools available for the coastline – coastal planning instruments, regulations, guidance, policy statements, etc. Coastal planners and managers are faced with the challenge of reconciling a number of conflicting and often incompatible demands, including economic development, tourism and recreation, urban expansion and agricultural development, while, at the same time, protecting areas of scenic or ecological value and vulnerable environments against the effects of erosion or flooding (Coccossis, 1985).

Islands and coastal ecosystems face serious dilemmas, such as: can the pressure for growth and development for tourism and leisure be permitted to destroy the very characteristics which make island and coastal areas attractive? Solutions to this question have tended to address specific issues, such as sustainability of the environment or of resources; the carrying capacity concept; conservation measures; planning and management schemes; policy instruments to absorb pressure for growth, etc. Among the above issues, the concept of sustainability and of carrying capacity are attracting considerable attention among researchers, especially in relation to island environments.

The proposed framework for the analysis and evaluation of the environmental impacts of tourism in coastal ecosystems attempts to incorporate the concept of carrying capacity, sustainable development and the so-called life cycle of a tourist product and belongs to the range of integrated approaches to the analysis of tourism/environment interactions which have long been proposed in the literature and applied in diverse circumstances (Coccossis and Nijkamp, 1995).

Patterns of Change in Coastal Areas

The changes in coastal areas as a result of tourist and leisure activities over many years reflect the development of local, national and international economies and trends in public knowledge and opinion. The intimate connection between the coastal fringe and its dynamic and interdependent zones (Fig. 10.1), including the aquatic–coastal–terrestrial elements of the system of coastal zones, has not always been recognized or respected in decision-making. It is well known that coastal areas have a variety of interrelated strategic functions including:

- ecological resource;
- recreational and tourist resource;
- a major part of the urban fabric of towns, cities and villages, especially those with traditional values;
- a transport route for passengers, freight and wildlife populations;
- military activity.

Because of this variety of functions and elements, there are numerous different organizations with responsibilities for managing and advising on these functions in any coastal system, as well as the many different users of the coast, some of whom have potentially conflicting views on how it should be managed. In the past, the coastal area was used as a major transport route by interested bodies, such as port, industry and transport authorities, as an area for urban expansion and, more recently, as areas for relaxation, recreation and tourism. Developments and changes in industrial, military and transport demands in the latter part of the present century have transformed the primary role of coastal areas to become a recreational resource, used not only by the local people who live in the vicinity of the system but by the vast number of tourists who visit the area.

However, in the case of recreation and tourism, growth in such vulnerable areas has been *ad hoc*, with sporadic planning, organizational and managerial advice from the various organizations involved. Obviously there is not an endless supply of space either in or around coastal areas. The dilemma for those managing these areas and their surrounding corridors is how to encourage access while conserving its natural heritage. One of the key functions of a coastal area planning and management strategy must be to coordinate the actions of the numerous organizations who are involved with its 'management'. Coordinating the needs and views of all of these 'actors', each with his own vested interests, is not an easy task, but one which is very important if a consensus is to be achieved.

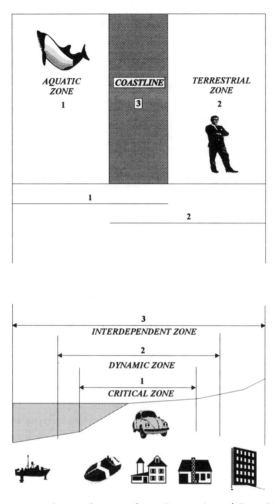

Fig. 10.1. The system of coastal zones (from Coccossis and Parpairis, 1993).

The growth in popularity of outdoor recreation and tourism has led to concern about the detrimental effect which visitors can have on the entire environment, both natural and built. In fact, parallel to this growth, there has also been an increase in public concern about the environment and the need to protect it. As a result, the topical issue of sustainability is now being introduced in recreation and tourism, as in all human actions, in an effort to protect the environment, or at least to limit human activity which may cause irreversible, detrimental change to the resources upon which tourism depends. In the long term the most effective way of achieving a sustainable level of recreational use will be by capitalizing on this concern for the environment and positively influencing behaviour. Although recreation might be an important use for some coastal areas, there are also other economic activities which are attracted to coastal sites, competing for the use of coastal resources and space. The coastal system is a natural resource and an important wildlife habitat. Its terrestrial part is also home to many people if well planned and sensitively managed. Recreational/tourist activities can exist in harmony with wildlife, as can local residents with other coastal users, while maintaining the coast as a valuable landscape feature, which is too often taken for granted.

As already pointed out, an additional feature of the coastal system is the number of organizations and interest groups with responsibilities for, or an interest in, its management. The role of all 'actors' is not always clear and there is sometimes confusion over both geographical and functional responsibilities. Indeed, the various topics currently being explored through research require a compromise between environmental constraints and economic growth. The major issues include: balancing ecological and environmental resources; introducing coevolutionary development (Nijkamp *et al.*, 1991); establishing complementary relations between economic development and environment; developing economic and non-economic system functions; and searching for ways to balance the interests of conservation and development.

Sustainability of Tourism Development in Coastal Areas

The concept of sustainable development has been interpreted in various ways, ranging from no development to compensation for heavy and pollutant development (Vernicos, 1987; WCED, 1987). Achieving sustainable recreation and tourism means finding a balance between conservation of the ecological resources – natural habitat and landscape features – and the use of these resources for certain purposes (Coccossis and Parpairis, 1992). The recreational use of coastal areas in particular is long established and, over the years, this use has brought substantial change, especially to the natural environment. The role of conservation is to protect and enhance the natural habitat of the coastal system and to ensure that new development in or near the system is carried out with a certain degree of sensitivity. However, specific sites along the coastal ecosystem require special designations which identify them as important wildlife habitats which should be safeguarded. General access to these and other conservation areas in coastal systems should be controlled in order to protect them against damage to the sensitive habitat. As access is important in offering opportunities for recreation and tourism in the coastal system, there is potential for conflict. In environmentally sensitive areas, the answer is to provide managed, informed access, ensuring that visitors are aware of when they can visit, on what scale, for how long, how they should behave and, of course, why the area is sensitive. Obviously, a process of public education would play a major role in managing access and influencing behaviour. This could take the form of a wide range of interpretative facilities, including information boards, guided walks, leaflets, school visits and working through local interest groups and educational trusts at local, national and international level.

Planning and policy-making for tourism development have been heavily concerned with the goal of obtaining a balanced relationship between tourism and the environment. The most important issues which stand out in this respect are: assessment of an area's carrying capacity and especially of the limiting factors determining the extent of tourism growth; well-structured planning approaches, ensuring balanced and sustainable tourism development and creating suitable policies for implementing the prescribed planning measures. The latter borrow elements from the broader class of environmental and development policies and adapt them to tourism. Their core concerns are: control of tourism growth away from environmentally sensitive areas; restrictions imposed on the types, extent and intensity of activities permitted in an area; proper management of waste generated by tourism; and the minimization of conflicts between tourism and competing land uses.

Methods of restricting the utilization of resources have been discussed over a considerable number of years, and nowadays they are frequently included in tourist resource management policies. In the future, the scope of such restrictions will probably expand in response to qualitative variations in the world tourist market, expressed in the noticeable change towards more active and special-interest holidays. This is a long-term process which involves, and will continue to involve, tourist sites based on natural, historical and cultural tourist attractions. The problem of restricting the utilization of resources for tourism, imposed by sustainable development goals, arises from the implications of the long-established concept of 'the site life cycle'. The popularity of each site is marked by ups and downs, depending on the inclinations of the dominant psychographical groups of tourists. This concept is similar to that referring to the product life cycle (Butler, 1980; Parpairis, 1993). However, while a product can be quickly adapted or completely removed from the market and replaced by a new technological solution, this cannot be so easily done with the tourist site as a market product. If its maturing is not detected early enough, it can lose its attractiveness both for a mass market and for possible market innovations.

Mediterranean tourism is now faced with the problem of environmental degradation, especially where tourism is based on the exploitation of natural resources. On the other hand, the development of mass tourism is too recent to produce examples of functional adaptation of mass tourism sites. Such adaptations can be put into practice in resorts which have been developed for a long time as, for example, on the Costa Brava, the Rimini Riviera or Mykonos in Greece. In theory, it is possible to extend the life cycle of a site if it is precisely planned in advance – following a course of action in consonance with the concept of sustainable tourism and using policy instruments such as carrying-capacity limits. This basic requirement can be imposed through integrated tourism planning and management, to provide a framework for site development in such a way that it can face and respond to changes in the tourist market and the environment.

Carrying Capacity of Tourism Development in Coastal Areas

The concept of carrying capacity in tourism has been approached from several perspectives, including biological and ecological, sociological, physical, behavioural, planning and design, and policy. The biological and ecological impacts of tourism have been studied in specific environments – islands, coastal zones, alpine areas, national parks, etc. From a behavioural point of view, visitor satisfaction with an area's environment and residents' perception of tourists have received scholarly attention (Vernicos, 1987; Parpairis, 1992). This viewpoint relates indirectly to the concept of the social carrying capacity of an area, i.e. the amount of social disruption beyond which both visitors and the local population experience negative consequences. However, a more wide-ranging viewpoint is required for the assessment of the environmental carrying capacity, which is necessary to plan for tourism development in harmony with the environment (Coccossis and Parpairis, 1992). In recent years, the issue of environmental carrying capacity has featured more frequently in development plans, strategic development projects and the wider planning and management literature in general. This is indicative of the increasing emphasis on environmental issues in planning and management studies and practices. More particularly, it is closely allied to the concept of sustainable development already mentioned, through the idea that there are identifiable capacity limits which should not be breached, in the interests of both present and future generations (Coccossis and Parpairis, 1991; 1993).

Environmental carrying capacity implies that the permissible amount of development – in this case, tourist development – can be defined by the state of the environmental limits and requirements. The underlying idea and the key to its political seductiveness are that the environmental carrying capacity of an area can be determined objectively. The term also implies that there are objective, immutable, unchallengeable environmental criteria and indicators which can be combined to define an area's carrying capacity – its maximum acceptable level of development. Such criteria emanate essentially from scientific areas like ecology, biology, planning, physics, chemistry and medicine or other disciplines in which principles are seen as inexorable. Moreover, scientific principles and laws are, in a way, permanently true, rather than being subject to changing academic fashions, political climate and planning practices. Environmental carrying capacity, once determined, will at least stand for much longer than other more transient decision criteria. Hence, the argument runs, if a conflict between those in favour of an ecology-orientated carrying-capacity approach and those in favour of development and growth arises, environmental and conservational considerations should prevail. This does not imply that carrying-capacity limits should be absolute and overrule social, cultural and economic considerations; rather, such limits should be dynamic in nature during the course of development, be flexible and, even more important, correspond closely to the needs and peculiarities of a specific area. In the case-study which follows, the concept of carrying capacity has been seen as a practical approach to certain problems of tourist development.

Environmental carrying capacity is best considered in a development plan, primarily at a local level, but also looking downwards to the specific site/project level and upwards to the regional scale. However, until our understanding of the interactions between the environment and development – human actions – is much more profound, the concept of carrying capacity cannot be used in planning and practice as an absolute tool offering exact measurements but, instead, as one which is under continuous revision, development and research. The range of options available to measure environmental carrying capacity include: a check-list; an inventory; guiding criteria and explicit environmental – as opposed to development – standards which depend on geographical and time scales; perception studies; and consistency issues and values, such as environmental relativity and quality evidence. These can be considered demanding requirements, which are not easily met for even the most general of environmental appraisals, but they must be subjected to scrutiny and research. Concepts, assumptions, methods and terminology should be clearly defined, so that the actual analysis of development impacts on the environment will be rational, systematic, explicit, technically sound and clear to all 'actors' involved in planning and management.

The above recommendations suggest some general policy implications for tourism planning and management. In the case of coastal ecosystems, the evaluation of carrying capacity can be valuable in providing general guidelines, which can be refined in time and space through trial and error. Local authorities should carefully formulate more detailed policies orientated towards development control, through planning and management, with the aim of achieving optimum capacity without saturation. Such policies should define the maximum capacity of the social and natural environment, the additional capacity made possible by overcoming bottlenecks and restraints – development corridors, transport centres, distribution in time and space, etc. – and, at the same time, the reduction of demand on certain sites, such as coastal areas, where capacity restraints cannot be overcome. A comprehensive plan for tourist development must foresee the provision of infrastructure and superstructure at acceptable standards; define norms, incentives and disincentives and a more coherent framework for the private sector; and introduce measures to increase the attractiveness of the coastal areas and to secure, at the same time, the financial means to maintain the coastal system. Finally, in order to protect the coastal environment in the interests of long-term touristic success, future research should address the need for a more holistic and systematic approach to the identification of critical zones and the assessment of potential environmental impacts resulting from the restructuring of the tourist product and from the expansion of the summer holiday-home phenomenon, a trend which, in the long run, can cause environmental degradation in coastal ecosystems.

Tourism has already attracted considerable attention in issues related to the development of small islands. The growing demand for tourism provides opportunities even for small and isolated places, by contributing to income and employment for islanders, but, at the same time, it generates direct effects on their sensitive and unique environmental resources (Coccossis, 1987). Such problems have already influenced local awareness of environmental issues, especially in those cases where the expansion of tourism has been rapid and intensive. The development of tourism gravitates to the island's natural features – landscape, plant and animal life – complemented by cultural and social attractions. Although these features and attractions should be protected, unfortunately tourism development can come into direct conflict with the protection of uniqueness, since it implies modernization, cultural change, urbanization and the extensive exploitation of resources. In many places, landscapes and coastlines are being visibly affected by widespread urbanization and the intrusive impacts of hotel, marina, airport and road construction. The consequent environmental degradation due to uncontrolled and intensive tourism development, however, inevitably affects tourism itself and the sensitive island ecosystem on which tourism is based. Special emphasis should be placed on the management of key island resources – coasts, fresh water, agricultural land, marine resources – in recognition of the interdependence of socioeconomic phenomena and environmental processes.

Mykonos Case-study

Mykonos is an internationally well-known tourist island, which has a high-quality environment and has experienced rapid tourist development during the last 30 years (Fig. 10.2). As it is a small island, it provides ample evidence of the impacts of tourism in coastal areas and the question of carrying capacity in a context of growth constraints. The historic evolution of

Fig. 10.2. Cyclades region in the Aegean Sea.

Mykonos, its unconventional natural beauty and its extraordinary architecture are the basic reasons for its popularity. The urban and architectural quality of the traditional settlement, Hora (the main town), has astounded famous architects such as Le Corbusier and I.M. Pei, as well as numerous visitors and tourists.

Mykonos is part of the Cyclades archipelago and belongs to a subgroup composed of three islands – Mykonos, Delos and Rinia. The total surface area of the island is 103.5 km² and the seashore is 81.5 km long (Fig. 10.3). The topography is mainly mountainous. The natural environment of the island, typical of the Cyclades, is characterized by low vegetation due to the dry climate, lack of fresh water and poor soil conditions, while the terrestrial environment is very diverse in vegetation types and rich in flora and fauna. Rural activities have adapted over the centuries to the low capacity of the terrestrial ecosystem, and few agricultural products are cultivated. The

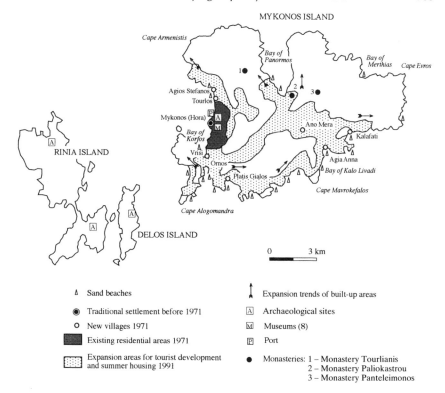

Fig. 10.3. Cultural resources and urban development on Mykonos island.

marine environment is of outstanding beauty and still largely undisturbed. The main attractions of the island are its beaches and the unique architectural heritage of the built environment. Because of its size and the traditional activities developed, this small island did not become a commercial, administrative or political centre, nor did it reach a status of economic influence, as was the case in the neighbouring island of Syros or other Aegean islands like Rhodes, Samos, Lesvos and Chios. Instead, it became a universal cultural centre, mainly because of its historical/cultural resources. For this reason Mykonos is strongly dependent on its indigenous natural and human-made resources (Coccossis and Parpairis, 1993).

It is clear from the historical development of the island of Mykonos that this development was a consequence of its central geographical location, the abundance of its cultural resources and its proximity to Delos, the sacred island of antiquity.

Little is known about Mykonos's history before the Greek revolution of 1821. According to tradition, the first inhabitants were Kars, Egyptians,

Phoenicians and Cretans. Various sources have suggested that there were initially two towns, which later became one – *c.* 200 BC. Until the time of Alexander the Great, the island was poor due to its rocky surface. At that time the island's economy developed, largely thanks to the port of Delos – 166 BC – and a democratic regime which continued during Roman times until the year 88 BC, when Delos was destroyed. Mykonos was then conquered by a series of invaders, including the Byzantines, Franks, Venetians, Catalans and Ottomans. During the sixteenth and seventeenth centuries AD, Mykonos became extremely weak as a result of wars, the loss of economic influence, changes of regime, mass emigration and, finally, the plague of 1678. Between 1700 and the early nineteenth century, Mykonos underwent a period of economic development, due to shipping, piracy and commerce. Thus, in 1821, the island was able to provide a large fleet and many men to fight for the noble cause of the Greek Revolution. The revolution encouraged further economic growth based on the island's shipping activities with, as a result, a significant increase in population. Since the mid-nineteenth century, however, the island has been losing population due to emigration – to the nearby island of Syros and abroad. It is important to note that Mykonos was a pioneer in economic management and common ownership. During the period between the two world wars, tourism began to develop on the island for summer or archaeological visits. This new activity came to an abrupt end, however, with the outbreak of the Second World War. Finally, over the past 40 years, the island has been transformed into a worldwide tourist and cultural centre.

The most important cultural resources of the area are the archaeological treasures of Delos and Hora, a settlement located near the island's small castle, with strong traditional character and amphitheatrical morphology, and a port which is a lively hub of activity. The natural landscape of the island, with its terraces, its monasteries and many churches, its windmills and the towns' picturesque neighbourhoods, are just a few of the resources which contributed to the island's 'fame' and helped overcome the problems of economic and population decline which affected it and several other small islands after the Second World War.

It is also of interest to consider the influence of the island's environmental characteristics on the development of its economy and society. From prehistoric times, myths were based on the rocky soil and the inhabitants' continuous struggle to discover ways of surviving on an island which offered sun, wind, sea and vicinity to the holy place of Delos. The poor, rocky soils are reflected in low agricultural production. Scarce rainfall aggravates the problem of agriculture, given the limited water-supply. Gentle slopes were transformed into terraces for cultivation as in many other Mediterranean islands. The climate is temperate, with mild winters but strong winds throughout the year. The constricting environmental factors led the inhabitants to make wise use of local resources and the majority of them turned to maritime activities.

Development on the island followed the general pattern of the socioeconomic evolution of the surrounding areas, which, from medieval times, were centred on the town's castle. The town developed around the port in a complex maze of streets and alleys – always narrow for climatic reasons and defence purposes. The rest of the island was divided amongst the farmers, who had their individual houses and stables, named *horia*. A second settlement grew in a densely occupied farming area, the community known today as Ano Mera. New settlements have been created on the coast and around the main town as a result of tourist development in the past 30 years.

At the beginning of the seventeenth century, there were 2000 inhabitants on the island, a population which increased to approximately 3000 at the end of the century. After a period of stability, from 1821 onwards, Mykonos received a large number of refugees from the nearby islands. From then on, it embarked on a period of decline, due in part to the decline of the shipping industry in the Mediterranean, and by the Second World War the population had dropped to 2150 inhabitants. The growth of the island population was greater in the period after the war, when tourism boomed. Of course, the significance of population figures for small islands which have developed as tourist resorts is limited, as wide-ranging variations occur between seasons, and in the high season the number of tourists is often several times greater than that of local inhabitants.

The traditional town of Hora is built on flat land, unusual for an Aegean settlement. This creates a complex urban tissue with a uniform character, remarkable harmony of volumes and minimum public space. Strict regulations have been imposed to maintain the architectural character of the buildings. The town has expanded and sprawled, increasing its surface from 33 ha in 1960 to over 100 ha in 1990. It accounts for 76% of the population, while an additional 16% live in Ano Mera. Apart from Ano Mera, rapid development has taken place in the Agios Stefanos/Hora/Ornos triangle in the southwestern part of the island. Expansion in this area has occurred in a haphazard way, and the site and extent of new growth has threatened the architectural and natural heritage.

Traditionally the economic activity of the island was based on agriculture, fishing and sea trade, but over the past 30 years it has changed dramatically to the visitor industry, involving the tourist and service sectors. Today tourism is regarded as the island's major export industry. Agriculture still survives, despite limited production because of harsh weather conditions, employment losses to other activities and the scarcity of water, which is largely consumed by tourism. The economic activities of Mykonos can be grouped into four sectors, which are, in order of importance: tourism, marine resources, construction and land speculation. Although tourism is the major industry of the island and offers considerable potential, excessive dependence on tourism over the last 20 years has made the island economy vulnerable to regional competition for tourism and fluctuations in the world tourist industry. Furthermore such 'monoculture' is incompatible with ecological sustainability (McElroy, 1990).

The first phase of tourism development on the island dates from the 1960s, with the creation of the first hotel, with 91 beds. This was the first of a series of hotels built throughout the country – especially in places with a strong historic/cultural image like Mykonos – in an early pilot programme promoted by the National Tourist Organization of Greece. This pioneer effort quickly stimulated the construction of more hotels and the provision of other types of accommodation by the private sector. Accommodation capacity jumped from almost 100 hotel beds in 1961 to 680 in 1971 and further to approximately 1800 in 1981. During the following decade, however, more hotel beds were added to the island's capacity. Estimates for 1991 placed total accommodation at 4724 hotel beds, with an additional 4000 beds in the form of 'rooms to let' – mainly in private houses; 960 beds in two organized 'tourist villages'; approximately 6000 beds in 3000 independent units of summer houses; and one camp-site with capacity for 504 visitors (Table 10.1). Official figures estimate the total capacity to be 16,214 beds for visitors. This figure, however, is believed to be a conservative calculation since, according to information provided by local residents and tour operators, total accommodation capacity for visitors exceeds 25,000 beds. The number of arrivals, in hotels alone, jumped from 5150 in 1965 to 22,294 in 1970, 36,150 in 1980 and more than 60,000 in 1990. Similarly, hotel bed-nights increased from 34,350 in 1965 to 82,650 in 1971, 206,299 in 1980 and more than 922,000 in 1990. These figures do not include a number of cruise ships which call in on the island on a daily basis, their visitors further crowding its narrow streets and other facilities.

The evolution in the demand for tourist services shows a continuous drop in the percentage of visitor arrivals at primary accommodation – hotels – in favour of arrivals at secondary forms of accommodation – rooms to let – a tendency which becomes even more marked in the case of bed-nights. Domestic tourism tends to prefer cheaper alternatives to hotel accommodation. It is worth noting that the rate of growth in the number of Greek tourists in terms of arrivals and bed-nights was several times greater in the last decade than the equivalent for foreigners. An overall view of the tourist activity on Mykonos is not complete without reference being made to the number of arrivals of one-day transfer visitors using Mykonos as an intermediate stop – 4000 daily. The local port and airport are the entry points: 30% of tourists and day-trippers arrive by air and 70% by boat. For the last 4 years, 80% of foreign and 90% of Greek tourists visit the island between May and September, with a peak during July and August, when foreigners account for 48% and Greeks for 52% of visitors.

Table 10.1. Tourist accommodation in Mykonos 1961–1991.

Type of accommodation	Accommodation						% change			
	1951	1961	1971	1981	1991	% of total	61/71	71/81	81/91	61/91
Hotels										
U	–	2	11	40	121	6.9	450.0	263.6	202.5	5,950
R	–	54	384	895	2,492	31.2	611.0	133.1	178.4	4,515
B	–	98	720	1,934	4,724	29.1	634.7	168.6	144.3	4,720
Rooms (NTOG)										
U	–	–	–	–	460	26.2	–	–	–	–
R	–	–	–	–	1,861	23.3	–	–	–	–
B	–	–	–	–	4,026	24.8	–	–	–	–
Camp-sites										
U	–	–	–	1	2	0.2	–	–	100	–
R	–	–	–	56	168	2.1	–	–	200	–
B	–	–	–	168	504	3.1	–	–	200	–
Villages										
U	–	–	–	100	172	9.8	–	–	72	–
R	–	–	–	200	480	6.0	–	–	140	–
B	–	–	–	800	960	5.9	–	–	20	–
Summer houses										
U	–	–	–	–	1,000	56.9	–	–	–	–
R	–	–	–	–	3,000	37.4	–	–	–	–
B	–	–	–	–	6,000	37.1	–	–	–	–
Total										
U	–	2	11	141	1,755	100	450.0	1,181.8	1,145.7	87,650
R	–	54	384	1,151	8,001	100	611.0	133.1	595.1	14,417
B	–	98	720	2,902	16,214	100	634.7	168.6	458.7	16,445

U, units; R, rooms; B, beds.
NTOG, National Tourism Organization, Greece.

Parallel to the expansion of the tourist industry, the island's population has also increased in size, in contrast to other islands, which have lost population. In 1951 there were only 2690 inhabitants, 2872 in 1961 and 3234 in 1971. During the following two decades, coinciding with rapid tourism development, the population increased even more rapidly, reaching 4850 in 1981 and over 6500 in 1991 (Table 10.2). It is estimated (Parpairis, 1993) that more than 6000 persons are directly or indirectly involved in the tourist sector, a ratio of one employee for every four beds. Growth in tourism demand was followed by an expansion of infrastructure through the enlargement of the port in 1978, the expansion of the small airport in 1990, the gradual improvement of the road network, and the construction of a surface dam in 1993, etc. These investments have further boosted the island's capacity to accommodate tourists and other visitors. Growing population and rising aspirations make it necessary for the productivity of economic activities to increase but, in the long run, small-scale, diversified and carefully managed resource-based enterprises owned by local residents may provide more benefits for the island's population than those promised by mass tourism organizations (McElroy, 1990). In the case of Mykonos, such a trend in the structure of the tourist sector can already be observed.

The island already presents some symptoms of saturation, and undesirable effects on the island's sensitive environments are emerging. These include congestion, lack of parking space, insecurity and water and soil pollution, especially during the peak summer season. Evidence shows that the limited natural resources of the island are insufficient to cope with the competing demands placed on these resources as a result of uncontrolled tourism development. The increasing number of residents and tourists is expected to use such large amounts of water that further investment in expensive desalination plants, water transport and/or the construction of water reservoirs may become necessary.

A large proportion of the island's extremely limited land surface has either been absorbed by intensive housing construction, tourism development and its accompanying infrastructure, or left unused for future speculation, thus causing widespread loss of agricultural land (Fig. 10.3). The two traditional settlements on the island, Hora and Ano Mera, together with other newly developed villages throughout the island – Agios Stefanos, Tourlos, Ornos, Vrisi, Platy Gialos, Agia Anna, Kalafati – on which the tourist industry was based mainly during the pioneer phase of development, have already been transformed in scale, volume of built-up areas, character and environmental quality as a result of the uncontrolled and rapid development of tourism (Figs 10.4 and 10.5). The value of the main town's architectural heritage and the threats from tourism development led, in 1973, to the protection of the area from future large-scale development through architectural and planning controls. However, effective conservation measures were absent during the 1980s, at the peak of tourist development, when the town

Table 10.2. Population change in Mykonos 1940–1991 (from National Statistical Service of Greece).

Residential units	Population						% change				
	1940	1951	1961	1971	1981	1991	40/51	51/61	61/71	71/81	81/91
Hora Mykonos (municipality)	1949	2535	2797	3009	4469		30.0	10.4	7.6	48.5	
Ano Mera (community)	547	256	836	629	680		−53.2	226.6	−24.8	8.1	
Settlements of the community of Mykonos	54	155	85	225	381		187.0	−45.2	164.7	69.3	
Total: Mykonos Island	2150	3546	3718	3863	5530	8500*	64.9	4.9	3.9	43.2	53.7

*Unofficial estimation of the 1991 enrolment.

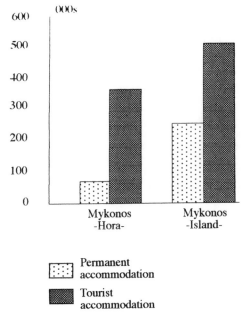

Fig. 10.4. Buildings for permanent residence and tourist use in Mykonos, 1991 (from Parpairis, 1993).

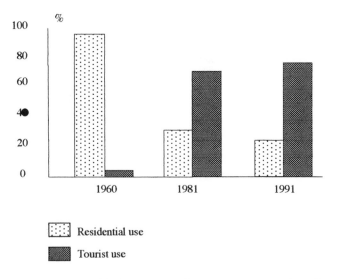

Fig. 10.5. Changes in use of buildings in Mykonos, 1960–1991 (from Parpairis, 1993).

expanded substantially and several new sites emerged as tourist nuclei (Fig. 10.6). Rapid tourist development, particularly of second homes, and its expansion on a scale which can be considered to be large in relation to the size and population of the island (Figs 10.7 and 10.8) not only threaten its rich built-environment heritage and the natural environment, but have also altered the socioeconomic structure and local culture.

Mykonos is being increasingly challenged by the large numbers of tourists visiting the island. Achieving the goal of ecologically sustainable economic development requires the adoption of effective environmental management strategies as part of a broader-based development policy. The situation on Mykonos suggests that there is a relationship between growth of tourism and the emergence of conflicts related to the island's scarce resources – productive land, underground aquifer, marine resources, infrastructure, etc. Empirical evidence from other tourism destinations, such as Italy, Malta, Croatia, suggests similar experiences, particularly in early-developed tourist resorts (Dragicevic, 1990; Young, 1991; Costa, 1993). Certain indicators can be introduced, even in a segmented approach, in the context of the island's carrying capacity, to express tourism/environment relations. These indicators include: the ratio of tourists to local population; the ratio of residential buildings to tourist buildings; the ratio of historical village stock to recently built areas; changes in building use and in land use through time, etc. (Figs 10.4–10.8). Perceptual studies can provide data regarding

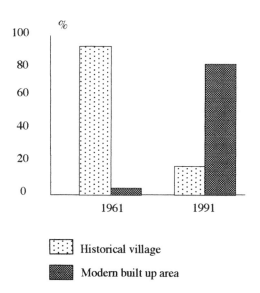

Fig. 10.6. Historical village and modern built-up areas in Mykonos, 1961–1991 (from Parpairis, 1993).

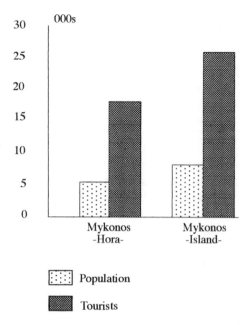

Fig. 10.7. Resident population and tourists in Mykonos, 1991 (from Parpairis, 1993).

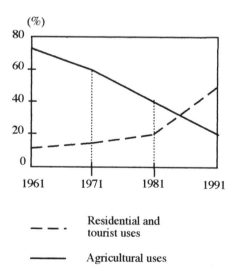

Fig. 10.8. Land-use change in Mykonos, 1961–1991 (from Parpairis, 1993).

satisfaction levels, which seem to have declined for both tourists and residents in the recent period of rapid tourist growth (Fig. 10.9). Construction has brought changes to the built and natural environment, not only in the form of aesthetic pollution but also of loss of quality according to human perception values. Congestion of traffic and visitors, litter and other waste, noise, etc. threaten the rich social activity, which is a major tourist attraction. Local relations have been altered to incorporate more formal ways of social exchange. Local tradition, values and models have changed with modernization but at a much faster rate due to tourism. Relations between tourists and locals have also changed. As a result of the 'cosmopolitan' type of tourism prevalent in the main town, tourists try to approach traditional hospitality, spontaneity, honesty of relationships and the authenticity of the experience in rural areas.

Conclusions

The above observations suggest certain public policy implications for tourism planning and management. Mykonos has already reached the crucial stage of tourism development, when the evaluation of carrying capacity can be valuable in providing general guidelines, which can be refined in time and space through trial and error. This example clearly demonstrates the need for local authorities to formulate carefully more detailed policies orientated towards development control through planning and management, with the aim of achieving optimum capacity without saturation. As already indicated, it is important to stress that such policies should define the maximum optimum capacity within the context of the social and natural environment, additional capacity generated by overcoming bottlenecks and restraints, and the reductions in demand necessary on sites where capacity restraints

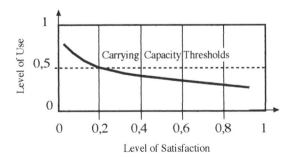

Fig. 10.9. Carrying capacity and personal satisfaction (from Coccossis and Parpairis, 1992).

cannot be overcome. Urgent application of such measures is required, based on land development monitoring, enforcement mechanisms, environmental education and public participation. The identification of the community's real needs, together with careful resource and technology assessment and option analyses may produce multiple benefits for the local population and for tourism. Priority should be given to measures directed towards increasing cultural tourism – by means of appropriate guidance, incentives and training – in order to escape from the production–consumption cycle and the model of mass tourism development. Emphasis should be placed on the redistribution of tourism demand and supply in space and time, on preservation policies for the natural, cultural and social environments, and on effective planning and management of the growth of both the permanent and visitor population. Finally, in order to protect the environment of the island of Mykonos, now approaching the stage of risk of irreparable damage, and in the interests of the long-term success of tourism, future research should address the need for a more holistic and systematic approach to the identification of critical factors and areas, and the assessment of potential environmental impacts as a result of the restructuring of the tourist product and from the expansion of the summer holiday-home phenomenon, already identified as a trend which could cause unforeseen and irreversible environmental degradation in the future.

References

Butler, R.W. (1980) The concept of a tourist area cycle of evolution: implications for management of resources. *Canadian Geographer* 24(1), 5–12.

Coccossis, H. (1985) Public policy for coastal areas. In: Hall, D.O., Myers, N. and Margaris, N.S. (eds) *Economics of Ecosystems Management*. W. Junk, Dordrecht, pp. 65–70.

Coccossis, H. (1987) Planning for islands. *Ekistics* 54(323/324), 84–87.

Coccossis, H. and Nijkamp, P. (1995) *Sustainable Tourist Development*. Avebury, London, 198 pp.

Coccossis, H. and Parpairis, A. (1991) The relationship between historical/cultural environment and tourism development. In: *Proceedings of the International Symposium on Architecture of Tourism in the Mediterranean, Istanbul*, Vol. 2, pp. 331–352.

Coccossis, H. and Parpairis, A. (1992) Some observations on the concept of carrying capacity. In: Briassoulis, H. and Van der Straaten, J. (eds) *Tourism and Environment: Regional, Economic and Policy Issues*. Kluwer Academic Publishers, Dordrecht, pp. 23–33.

Coccossis, H. and Parpairis, A. (1993) Environment and tourism issues: preservation of local identity and growth management: case study of Mykonos. In: Konsola, D. (ed.) *Culture, Environment and Regional Development*. Regional Development Institute, Athens, pp. 79–100.

Coccossis, H. and Parpairis, A. (1995) Assessing the interaction between heritage, environment and tourism: Mykonos, Greece. In: Coccossis, H. and Nijkamp, P. (eds) *Planning for Our Cultural Heritage*. Avebury, London, pp. 116–142.

Costa, P. (1993) Managing mass tourism in vulnerable art cities: Venice and other European destinations. In: Konsola, D. (ed.) *Culture, Environment and Regional Development*. Regional Development Institute, Athens, pp. 63–77.

DoE (Department of the Environment) (1993) *Coastal Planning and Management: A Review*. HMSO, London, 178 pp.

Dragicevic, M. (1990) *Methodological Framework of Assessing Tourist Carrying Capacity in Mediterranean Coastal Zones*. PAP, Zagreb, 43 pp.

McElroy, L.Z. (1990) Challenges for sustainable development in small islands. In: Beller, W., Ayala, P.G. and Hein, P. (eds) *Sustainable Development and Environmental Management of Small Islands*. Parthenon Press/UNESCO, Paris, pp. 229–316.

Miossec, J.M. (1977) Un modèle de l'espace touristique. *L'Espace Géographique* 6(1), 41–48.

Nijkamp, P. Van den Bergh, C.J. and Soeteman, F. (1991) Regional sustainable development and natural resource use. In: *Proceedings of the World Bank Annual Conference on Development Economics, 1990*. IBRD/World Bank, Washington, DC, pp. 153–187.

Parpairis, A. (1992) The evolution of the life cycle of a tourist product. In: *Proceedings of the 3rd International Conference on Environmental Science and Technology, Lesvos, Greece*, Vol. B. University of the Aegean, Lesvos, pp. 673–689.

Parpairis, A. (1993) The concept of carrying capacity. PhD thesis, Department of Environmental Studies, University of the Aegean, Mytilene, Greece.

Pearce, D. (1989) *Tourist Development*, 2nd edn. Longman, Harlow, 326 pp.

Vernicos (1987) Three basic concepts: man as part of the environment, carrying capacity, conservation, some further consideration. In: Hall, D.O., Myers, N. and Margaris, N.S. (eds) *Economics of Ecosystem Management*. W. Junk, Dordrecht/Boston/Lancaster, pp. 41–46.

WCED (World Commission on the Environment and Development) (1987) *Our Common Future*. Oxford University Press, Oxford, 267 pp.

Young, B. (1991) At risk – the Mediterranean, the Mediterranean way and Mediterranean tourism: perspectives on change. In: *Proceedings of the International Symposium on Architecture of Tourism in the Mediterranean, Istanbul*, Vol. 2, pp. 43–48.

11 Sustaining Local Cultural Identity: *Social Unrest and Tourism in Corsica*

Introduction

Over the past 40 years, Corsica, like many other mainland or island destinations in the Mediterranean, has undergone deep-seated changes as a result of tourist activities. Although tourism is not the most important source of income on the island – unlike, for example, the Balearic Islands – and tourist developments exist on a much smaller scale, this is not due to the absence of grandiose projects for the promotion and development of tourism on the island. French government officials had, in fact, earmarked the island for development in the mid-1950s, having determined that it was 'an incomparable location' for European tourists, in particular, in search of the sun and warm sea. A series of plans for the development of tourism on the island were thus drawn up and put into effect by the Paris administration. These plans, however, were hotly contested and protested against by a sector of local public opinion and their elected representatives, and tourist developments became the target of terrorist attacks. There is no doubt that this unexpectedly violent reaction caused the island's potential development as a tourist destination to be drastically curtailed.

Indeed, this attitude of protest constitutes one of the distinguishing characteristics of the island and the development of tourism there in recent years. In search of its own blueprint for development, and subject to drastic social and spatial transformations, emotions on the island run high. Tourism, like other forms of development, is seen to be imposed from outside and, as such, should be rejected. Such an attitude, which must be understood within the wider context of the island's rejection of French government rule in Corsica, has played an important role in the social and political changes that have taken place on the island over the past 40 years.

At first, it took the form of a fairly generalized, though not always explicit, criticism of development in its different shapes and forms. Did Corsica have to – and if so why? – adopt the same norms, criteria and forms of development that existed in the industrialized world? Was there no alternative available whereby local organization and control of the means of production, Corsican participation in the development of their island and the establishment of new economic and cultural links could replace the concept of growth with the concept of development, thereby satisfying the specific demands and aspirations of the inhabitants of a region and, what's more, an island? (Richez and Richez-Battesti, 1982; 1986).

One might suppose that Corsica's opposition to the tourist industry, and in particular the unconventional manner in which that opposition was expressed within a democratic state, i.e. through terrorist attacks on tourist facilities, reflected the island's desire to participate in the decisions made concerning local development and to make that development compatible with local culture and values. It was an attempt to break with the prevailing economy of masses and to promote a form of development respectful, amongst other things, of cultural identity and plurality, controlled by local enterprises, reproduced by them and which, therefore, would be more easily sustainable.

The following hypotheses may be ventured as to why tourism should have acted as a catalyst for islanders' discontent. None is exclusive of any of the others.

- The first and most important reason may be found in the importance given to tourist activities by the French government in its plans for the development of the island. The way in which tourism was to be developed there, the type of tourism and the investments to be made, the sites to be developed, etc., i.e. the socioeconomic and political consequences of government proposals, allowed protest actions to assume a highly symbolic and exemplary value.
- A second would be that tourism was a relatively new activity and it was easier and more profitable to attempt to exert one's influence on an incipient process than to act upon structures already in place and in operation.
- A third would be that the organization of this sector seemed less rigid than that of other sectors; would this not be the result of the smaller size of the firms involved and the characteristics of the tourist market?
- A fourth would be that as far as many Corsicans were concerned, including those who did not question the development taking place on their island, tourism and leisure activities were not viewed as activities capable of producing income – like, for example, industry – or of promoting economic growth and the integration of the island's economy into the Western economy. It should be pointed out that similar opinions have also been expressed in other regions of France.

- A fifth would be that the link between tourism and foreigners is immediately obvious. Thus, although all other sectors of the Corsican economy, including the black economy, depend to different degrees on foreign markets, tourism serves to focus the protests of those who refuse to accept the type of economic development that underscores their dependence.

The socioeconomic changes taking place in Corsica today must be understood within the context of the island's history, i.e. within the context of its resistance to French rule, which would appear to be a constant in its history. They must also be understood within the wider context of macrosocial development, paying particular attention to the local situation, the interaction of the different policies adopted by the islanders or groups of islanders, the confrontations, conflicts, disorders and inequalities.

Opposition to tourism, in fact, symbolizes and reflects the principles of a certain sector of Corsicans who were protesting at the growth of the economy in general. It was not that they rejected the concurrence of activities on the island but that they refused to give up their ancestral rights over the island to give them over to 'the colonizing power of money'. In short, they refused to give up their inheritance, selling it off to foreigners who would make a profit out of it. The Corsicans' claim to the right to make their own decisions concerning the development of their island, and, for some, the right to manage its development within an integrated system, cannot be clearer. This does not, of course, mean that the decisions made locally are any better for the island than those that have been the object of protest.

Plans for the Development of Tourism within their Sociopolitical Context

When tourism began to develop in the 1950s, the socioeconomic situation on the island of Corsica was characteristically one of fragmentation and underpopulation. It should be noted that the population of the island was ageing and, numerically, small: 170,000 inhabitants in 1954, with a population density of 20 inhabitants per km². This low density of population varied, moreover, round the island. The reason for this may be found in two centuries of migration from rural areas and in 'exile'. A ready source of labour, the island's inhabitants satisfied the demand for recruits in government posts on the mainland or in the French army, on the continent or overseas. As French nationals, Corsicans were often educated in France. Those who were educated on the mainland were not only regarded as being the epitome of success and social advancement by those who had remained on the island in charge of its destinies, but also the main guarantors of the integration of this reputedly ferocious people into the French state.

When the French government decided to institute a policy of regional development to offset differences observed in the economic development of different regions in France, Corsica, naturally, was not forgotten. In 1957, the Plan for Regional Action (*Plan d'Action Régional*) (MAEF, 1957) in Corsica was put into effect less than 2 years after the basic guidelines had been established and set down in the Decree of 30 June 1955. The Plan for Regional Action instituted a plan for economic development and a programme of action to be implemented by the government. It envisaged the overall, harmonious development of all local potential and referred to tourism as the 'key to the renaissance of Corsica'. Tourism alone was 'capable of bringing about economic growth' and of providing the 'means to prosperity'. 'Corsica should take advantage of this opportunity to overhaul its economy' since this will allow her to 'possess a tourist location of international standing' (MAEF, 1957). According to the nationalist leader, Edmond Simeoni, the Plan marked

> a change of direction in French policy in Corsica: it was when France decided to change the status of the island from that of a colony that had been settled – where the only resource exploited was its manpower, as its inhabitants were forced into exile – to one which was to be resettled

since this new activity would attract labour from the mainland or from further afield.

Two companies were set up on the island: the *Société de Mise en Valeur de la Corse* for developing agriculture and the *Société d'Équipement Touristique de la Corse* for developing tourism facilities. The failure of the latter to fulfil its mission was complete: whilst it should have built 100 hotels, only four were completed, with a total of 234 rooms, for which, moreover, difficulty was encountered in finding buyers. The company's efforts to sell or promote the sale of land or private tourist complexes were more successful, as were its efforts to attract investment from private finance corporations. However, in one of its operations, in Pinia, Ghisunaccia, where 30,000 beds were planned, the company ran into difficulties, causing it to reduce its activities, before finally disappearing in 1977.

Fifteen years after Corsica's first regional plan of action, while second homes were beginning to be built along the coastline and numerous plots of land were being put up for sale, the 1971 Plan for the Development of Corsica – *Schéma d'Aménagement de la Corse* – gave tourism on the island even greater importance than the first (MIAC/PRC, 1972). It proposed a 10-year programme to provide the infrastructure necessary for the development of tourism and, according to the authors of the plan, 'would not affect the acquisition of land [by promoters] as this had probably been carried out a long time before'. The *Schéma* formed part of the Sixth Development Plan, which was devised to continue the industrialization of France begun under the Fifth Plan and to develop the larger regions according to their potential or to the role they had been assigned within the Plan. Thus, the Report of the Commission on Tourism in the Sixth Plan reiterated the government's desire to develop the 'location' of Corsica. The number of visitors to the island was to increase from an estimated 360,000 tourists in 1970 to 2.2 million – maximum estimated – in 1985, by increasing accommodation 2.5 times and the number of tourists 6.1 times.

The number of private construction companies and developers multiplied: Portu Monaghi – 1500 beds; Tuara – 3500 beds; Capendula – approx. 5000 beds; Côte de Diane – 6600 beds; Portu Giraglia – 10,000 beds; Pinia – 30,000 beds; Testa Ventilegne – 35,000 to 100,000 beds; etc. (Fig. 11.1).

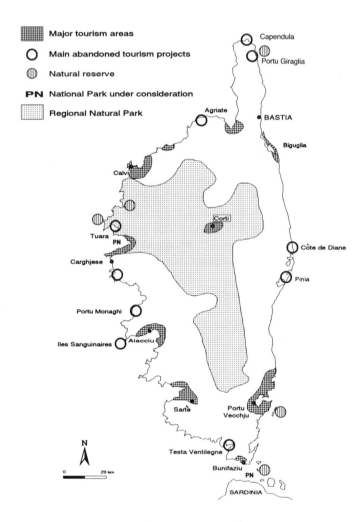

Fig. 11.1. Some aspects of tourism in Corsica.

Faced with this huge increase, environmentalists, supporters of regional autonomy – who first appeared in 1967 – and, subsequently, Corsican nationalists – who appeared in 1972 and 1973 but more especially in 1976, with a schism occurring in 1990 – all warned islanders of the dangers inherent for Corsica and the Corsicans in participating in such a large-scale increase of tourism, which would become unmanageable and which they would be unable to control.

These arguments, which won the support of a sizeable sector of public opinion as well as that of several of the island's elected representatives, together with the development of events and the onset of violent actions, led to the failure of most of the projects. If tourism had developed in Corsica as the government had planned it to, the island would have provided tourist facilities similar to those available today on the Balearic Islands, only on an island twice the size, with a fragmented infrastructure and a population in 1982 of 240,000 inhabitants, i.e. 2.5 times smaller than the Balearic Islands. The comparison with the Balearic Islands is often made in Corsica and the mainland since the Balearics epitomize the negative effects of mass tourism – referred to pejoratively in French as *la Baléarisation*. It should be pointed out, however, that the total surface area of the Islands is 5014 km^2, with a population of 768,000 inhabitants in 1990. A total of 6–7 million tourists per year visit the Islands, where the gross domestic product (GDP) is the highest in Spain and other Mediterranean islands. These data are interesting, particularly in view of the harsh living conditions and limited means of production which existed on Mediterranean islands up to the middle of this century or when they no longer belonged to a rich nation (Richez and Richez-Battesti, 1982).

The Questioning of the 'All Out for Tourism' Policy, 1970–1980

A number of Corsicans objected in general to the idea of a seasonal invasion of their island by 2 million or so tourists and, as might be expected, refused to accept the outlandish projects proposed, justifying their arguments on the basis of the demographic characteristics and socioeconomic situation of the island. Indeed, they were quick to realize that the fabric of the island was too weak and fragile to withstand such an influx of tourists without running the risk of possible impairment and subsequent disappearance. The very process of creating the infrastructure necessary for tourism was even questioned; the island only served as the backdrop to an economic activity the Corsicans would have no control over and in which they believed they would find only second-rate employment, if they were lucky.

It should be pointed out that, while these projects were being developed, the sociopolitical and economic situation in Corsica continued to deteriorate and it did not seem that the 'all out for tourism' campaign could do much to change the situation. Similar in concept to the programme of redevelopment carried out in the agricultural sector, in particular, on the Eastern Plain, the Plan could only accelerate the exploitation of the island at the expense of its inhabitants. The analysis and arguments put forward by nationalists and supporters of regional government were clear on this point. Their opinions were not shared, however, by the government in Paris, presumably because they were not voiced by the islanders' elected representatives and were thus not 'legally' representative of their opinions. The authorities' refusal to listen and to take necessary action was met with terrorist action. Hotels that were under construction, such as the Hotel Club de la Marana; holiday villages such as Biguglia, both south of Bastia; holiday clubs such as Club Méditerranée in Carghjese or the German-owned Safari Club; marinas such as Portu Monaghi, south of Carghjese; second homes, sometimes built without permission and in contravention of land occupation regulations, were either destroyed or badly damaged as a result of bomb attacks. These attacks, together with attacks on government and mainland-based institutions, such as banks, tax offices, police stations, the *Rectorat*, Ministry of Agriculture offices, etc., evidenced the islanders' determination to reject certain forms of exploitation of their homeland and their decision to halt the process of development before it was too late.

An analysis made of the situation between 1970 and 1980 caused the government to push forward as fast as possible with its projects for developing tourism. Ten years after the great agricultural reforms had been carried out on the Eastern Plain – in general, clumsily – it was easy to assess the results of the implementation of a grandiose 'neocolonial' agricultural project, which both benefited from state subsidies and had, under the guise of aid to Corsica, attracted most of the young Corsicans who had not qualified for the credits or subsidies they had applied for. Parallels have been established between these large agricultural exploitations, which were incorporated into the great wine-growing industry, and the management of holiday clubs such as the Club Méditerranée or hotels belonging to the Compagnie Générale Transatlantique, which, at the time, enjoyed the monopoly of ferry services between the island and the mainland. Islanders opposed these large holdings not only because of their staff recruitment policies, but also because they imported consumer goods almost exclusively from the mainland instead of turning to local supply and potential to satisfy specific demands. Those employed in the tourist industry were either highly qualified and came from the mainland, or were non-qualified and recruited from countries where labour was cheap, in abundant supply and more compliant: North Africa, Turkey, Yugoslavia, depending on the cases and the year. In 1982, despite the changes brought about in employment policies in the tourist industry, local labour represented only 40% of the total labour force –

generally not highly motivated by the possibility of employment in the catering sector.

As from August 1975, Corsica was the scene of a series of dramatic events, such as the Aleria and Bastelica/Fesch scandals; the trial and sentencing of approximately 100 Corsican devolutionists and nationalists to long prison sentences by the military courts; and the decision of another 100 or so Corsicans to join the resistance and continue their struggle against the authorities from within clandestine groups. During this period, slogans and inscriptions abounded, appearing on walls, billboards, traffic signs and electricity posts. Amongst them was the famous 'IFF' – *I Francesi Fora*, 'French go home'. Tourists, most of whom were ill-informed or uninformed and had come to the island to enjoy their holidays, could not and would not understand the motives behind the slogans or the violent episodes that took place on the island. They felt uneasy and unhappy at a situation they felt victims of, even though they had not been personally affected by events, and both the number of individuals and the flow of visitors fell between 1983 and 1984 (Richez and Richez-Battesti, 1986) (Table 11.1).

House sales were soon affected. According to the Department of Regional Development, the construction of approximately 2000 second homes per year in the years preceding 1975 dropped to hardly 500 in 1978. Not only were large-scale projects abandoned but so too was some construction work

Table 11.1. Evolution of the annual flow of tourists and the population in Corsica, 1955–1990 (from INSEE, 1994).

		Number of bednights		
Year	Number of tourists (thousands)	(millions)	(average annual variation, %)	Population (thousands)
1955	50	–	–	170
1960	130	2	–	173
1965	240	8	60	185
1970	525	9	2.4	215
1975	820	14	11	230
1980	1150	19	7.2	235
1982	1270	26	18.5	240
1983	1170	23	−12	–
1984	1100	19	−21	–
1985	1250	21	11	–
1990	1750	21	0	250

NB: No data are available for the summer season (May–October) 1991–1994. There was a marked drop in the number of tourists on the island during this period: 100,000 fewer tourists (−6%) and the number of bednights decreased to 1,350,000 (−6%). In the summer of 1994, 1.5 million tourists visited the island.

already under way, and the construction industry found itself in difficulties after having enjoyed a period of marked expansion.

Other events occurred which, in conjunction with those directed against tourism, contributed to further worsen the socioeconomic situation on the island. In addition to bomb attacks on second homes, forest fires – almost certainly started deliberately – forced camp-sites and holiday camps to be evacuated, and there was a series of *nuits bleues*, during which dozens of bomb attacks took place on the island and on the mainland (Table 11.2). People from the mainland were pressurized and threatened, particularly members of the teaching profession living in Corsica. The nationalist movement became increasingly radical and split into two camps.

Nevertheless, despite this difficult situation, the flow of tourists during the decade under study continued to increase overall. Since 1985, Italian visitors have replaced north European and French tourists troubled by the atmosphere of violence on the island. By 'saving' several tourist seasons, they have helped disguise the problems on the island – the number of Italian visitors to the island rose from 18,000 in 1977 to 61,000 in 1981 and from 229,000 in 1987 to a maximum figure of 361,000 in 1992. Thereafter, the number of visitors was affected by the recession in Italy and the devaluation of the Italian lira.

Terrorist attacks directed against tourism, however, are not the only events to be taken into consideration when analysing the sociopolitical situation on the island – they account for only 15% of the attacks carried out between 1973 and 1985. A whole series of actions attributable to Mafia-like activities or to banditry perpetrated in the name of nationalism complicates the analysis and understanding of the situation. It is well known that in Corsica, like other places, establishments such as night clubs, bars, discothèques, etc. are associated with racketeering. This is a consequence, and

Table 11.2. Evolution of the terrorist attacks on tourism facilities 1973–1985 (based on statistics provided by Ministère de l'Intérieur).

Tourist facilities	Number	Percentage
Second home	275	38.7
Holiday camp/village	73	10.3
Hotel/hostel	60	8.5
Camping/caravanning site	15	2.1
Travel agency, tourist office, *syndicat d'initiative*	11	1.6
Bus, camping-car, caravan	18	2.5
Boat, aeroplane	35	4.9
Bar, restaurant, dance hall	223	31.4
Total	710	100

one of the more important effects on society, of the development of tourism on the island. A series of personal vendettas, as well as commercial and political rivalries, were settled by violent means. Corsica is not the only case – the departments of Var and the southern Alps are similar cases – where concern exists regarding tourist and leisure activities and certain forms of violence. This violence may be interpreted in different ways, depending upon whether those involved are supporters of central or autonomous government, groups of militant nationalists or antinationalists, Corsicans or non-Corsicans, and whether responsibility is claimed or not. The Mafia-like tendency of nationalism – characterized by the imposition of a 'revolutionary tax', the nationalists' form of racketeering – is also found in other nationalist organizations in Europe.

The appearance of drug abuse on the island, with the amount of money and envy it generates, has added to the violence and general state of confusion already existing on the island, much to the concern of the general public. Another problematic aspect is the damage done to environment as a result of forest fires, the scourge of Mediterranean countries. The origin of many of these is to be found in the Corsican custom of 'burning off the land'.

The level of violence present in Corsica is marked, and has evolved out of a situation unique to the island, which, on the one hand, brings to mind Sardinia and Sicily, and yet, on the other, clearly distinguishes it from Irish and Basque nationalist movements, both of which are associated with large numbers of violent deaths. Corsica is also quite different from the Balearic Islands from the point of view that violence is hardly a problem at all in the Balearic Islands and so maximum advantage can be taken of the flow of visitors to the islands.

Tourism as an Integral Part of the Island Economy

The situation described above, which has lasted since the beginning of the 1970s, has seriously curtailed local economic activities since potential investors have turned their back on the island, given the risks involved. Thus, there has been a marked slowing down of development projects. Almost all the major projects envisaged have been halted, as have a large number of smaller ones, thus placing the Construction and Public Works sector – an important employer of labour (Table 11.3) – in a difficult situation, which neither public works nor reconstruction schemes associated with bomb attacks can keep in business. The different effects of the advent of tourism have voluntarily been reduced to a minimum, out of fear of racketeering and bomb attacks. Even so, as we have seen, the flow of visitors to the island continues to increase and tourism, whether it is a good year or a bad year, has become an integral part of the economy in Corsica.

Table 11.3. Distribution in % of the active work force according to economic
activities (from INSEE, 1986, 1994).

Sectors	1962	1975	1982	1990
Agriculture and fishing	32.7	20.3	13.4	7.1
Industry	9.5	6.4	7.3	7.2
Construction and public works	13.0	19.4	15.5	10.5
Services	44.8	53.9	63.8	75.3
Total	100	100	100	100

Despite violent protests against the development of tourism, the tourism
industry does in fact constitute an important source of income on the island.
INSEE (1991) has estimated that in 1987 tourism, excluding external transport, contributed 33 million French francs to the island economy – 20% of
household income – and provided 10,000 seasonal jobs during the month
of August – the equivalent of 15% of the permanent jobs available on the
island. In 1994 tourism was expected to contribute 49 million francs. It also
accounts for an important part of the black economy.

According to INSEE (1994), 62% of household incomes come from
government sources – salaries in the public sector and social benefits. In
such circumstances, the islanders' response to the tourism sector – which
includes total rejection by some Corsicans – is easily understood, since
the socioeconomic situation on the island is such that there is no need
to develop its tourist potential. The islanders are content simply to live
off a tourism industry which is limited exclusively to the summer months
and easily attracts tourists, since the sun and sea are free, and, despite
local indifference, stubbornly increases in importance as the number of
visitors continues to rise.

This peculiar situation has largely determined the type of accommodation available on the island. Camp-sites doubled their capacity between
1980 and 1994, and constitute over 25% of all visitor bed-nights. Hotel
bed-nights, on the other hand, account for only 12% of the total, and
many hotels are encountering financial difficulties. Uncontrolled camping
in the countryside is still, by and large, accepted, despite its detrimental
effects on the environment – degradation of soil and coastal vegetation
and production of waste – while generating very poor economic returns.

Tourism: a Challenge for Future Sustainability

Over the past 25 years, despite the difficulties it has faced, Corsica has,
it would seem, managed to avoid the dangers of an 'all out for tourism'

policy aimed at developing only one economic activity on the island. This impression does not, however, represent the reality of the situation since, because other sectors of the economy have not been developed, tourism is, in fact, the only activity that exists.

Perhaps it is time now for islanders to meditate upon the choices available to them for development. They have, nevertheless, clearly stated that they want coherent development, despite their contradictions. Meanwhile, the violence – which is a constant challenge to the principles of the democratic state, the conflicts that arise – in particular involving transport to and from the island, thus holding the islanders effectively to ransom – and, finally, the small but active group of islanders that are determined to make the most of the troubled situation do nothing to make the choice any easier. The situation on the island is unique in the Mediterranean. Part of it has to do with the island's dependence on French government support and the help it has been given, but part of it has to do with the adverse effects this has given rise to.

The social situation on the island closely mirrors the economic situation. Corsican society is in the throes of a profound crisis associated with the evolutionary processes of the twentieth century. This has been exacerbated by the disappearance of traditional customs, the lowering of moral standards and the loss of basic human values, probably related to the emergence of various nationalist movements some 20 years earlier and to the situation of unrest which has existed since. Probably the only aspect of society which has witnessed renewal and significant development has been artistic expression, in the form of music and song, in particular, and also drama.

The positive aspect, and there is one, of the difficult socioeconomic situation which has affected Corsica since the war is the conservation of the natural environment. The Regional Natural Park has been enlarged and currently covers over one third of the total surface of the island, affecting 38% of the communes (Fig. 11.1). Four beautiful Nature Reserves have been constituted to preserve several stretches of coastline and a number of small islands from development. The creation of two National Parks is currently under consideration, one of which would span part of both Corsica and Sardinia. In addition, construction companies – *sociétés civiles immobilières* – and some private property owners, forgoing the idea of promoting tourist accommodation, have resold the land they bought at low prices to the *Conservatoire du Littoral*, the government agency responsible for purchasing land for conservation purposes and for the development of facilities for tourism and leisure activities. This agency has been active in acquiring the land private investors have been selling, for the benefit of the public at large. By January 1993, the *Conservatoire* possessed 10,300 ha, including 122 km of officially protected coastal fringe and 15% of the entire coastline of the island.

The conservation of nature is essential if sustainable tourism is to be developed. There can be no doubt that the natural landscape is Corsica's greatest asset and that of whatever type of tourism it wishes to develop. The socioeconomic problems of the island, however, have still to be solved, and this will be no easy task. A marked improvement in the economy and greater understanding of the specific needs of the island are necessary if the obstacles to its development are to be overcome. A change in moral values is required to regenerate public life, bring about peace and promote a strong, dynamic island society.

References and Further Reading

Buckley, P.J. and Klemm, M. (1993) The decline of tourism in Northern Ireland: the causes. *Tourism Management* 14(3), 184–194.

Delors, J.P. and Muracciole, S. (1978) *Corse: la poudrière*. Éditions Alain Moreau, Paris, 380 pp.

Dressler-Holohan, W. (1981) *Développement économique et mouvement autonomiste: le cas de la Corse*. Rapport CORDES, Paris, 300 pp., plus annexes.

Helios Consultants (1987a) *Étude tourisme et transport 1987*. Chambre de Commerce et d'Industrie d'Ajaccio-Sartene, Aiacciu, 67 pp.

Helios Consultants (1987b) *Clientèles touristiques et saisonnalité*. Chambre de Commerce et d'Industrie d'Ajaccio-Sartene, Aiacciu, 87 pp.

Helios Consultants (1991) *Quinze années de tourisme en Corse. Evolutions et mutations, 1977–1990*. Chambre de Commerce et d'Industrie d'Ajaccio-Sartene, Aiacciu, unnumbered.

INSEE (Institut National de la Statistique et des Enquêtes Économiques) (1976–1995) *Économie corse*, periodical publication (five numbers per year).

INSEE (1986) *Tableaux de l'économie corse*. INSEE, Aiacciu, 146 pp.

INSEE (1991) *Tableaux de l'économie corse*. INSEE, Aiacciu, 149 pp.

INSEE (1994) *Tableaux de l'économie corse*. INSEE, Aiacciu, 153 pp.

MAEF (Ministère des Affaires Économiques et Financières) (1957) Approbation du Programme d'Action Régionale pour la Corse. *Journal Officiel* 19 d'avril, 4136–4145.

MIAC/PRC (Mission Interministerielle pour l'Aménagement de la Corse and Préfecture de la Région Corse) (1972) *Schéma d'aménagement de la Corse. Travaux et recherches de prospective*. La Documentation Française, Paris, 96 pp.

Richez, G. (1981) Le développement du tourisme à Mallorca (Baléares) et ses conséquences socio-économiques. *Cahiers de la Maison de la Méditerranée* 4, 73–108.

Richez, G. and Richez-Battesti, J. (1982) Tourisme et mutations socio-économiques en Corse et à Mallorca. *Études Corses* 18–19, 329–364.

Richez, G. and Richez-Battesti, J. (1986) La contestation du tourisme en Corse et ses implications économiques et sociales. *Études Corses* 26, 49–96.

Richez, G. and Richez-Battesti, J. (1992) Tourisme en espace rural dans l'île de Corse. *Bulletin de la Société des Sciences Historiques et Naturelles de la Corse* 662, 17–37.

Ryan, C. (1993) Crime, violence, terrorism and tourism: an accidental or intrinsic relationship? *Tourism Management* 14(3), 173–183.

12 European Perspectives on Sustainable Tourism

J. Arwel Edwards and Gerda K. Priestley

Issues Raised

The ten case-studies presented in this book have concentrated on two of the principal tourism environments which experience high levels of demand and in which the interaction between tourism and the environment is important: namely, protected natural/rural areas – hereafter referred to as natural areas – and popular coastal destinations. A number of wide-ranging issues set in different contexts have been raised, and three of these are of particular importance: setting, scale and the present level of development. Table 12.1 summarizes the themes covered and the conflicts resulting from tourism development. The geographical settings range from natural, through partially modified rural, to built environments; indeed, in many cases, the environments combine elements of both. In fact, only one (Loch Lomond, Chapter 2) emerges as a totally natural environment, a reflection of the wide range of European tourism environments.

The areas studied vary in size from small, highly localized sites to entire regions. Moreover, the transformations as a result of tourism development occur at very different scales of intensity. Obviously the impacts of tourism along the Catalan and Valencian coastlines (Chapters 7 and 8) are more widespread than on the island of Mykonos (Chapter 10) or the individual marinas studied in the United Kingdom (Chapter 6). In some instances, there have been issues raised at a local level, such as along the shores of Loch Lomond and, in others, the problems affect a much wider area such as the Adriatic Sea (Chapter 9) where the impacts generated in one place can have wide-ranging spatial repercussions elsewhere. In fact, it is known that the scale of tourism impacts can extend to continental and worldwide dimensions.

© 1996 CAB INTERNATIONAL. *Sustainable Tourism? European Experiences*
(eds G.K. Priestley, J.A. Edwards and H. Coccossis)

Table 12.1. Characteristics of the case-studies.

	Chapter									
	2	3	4	5	6	7	8	9	10	11
	Lomond	Gower	Forest	Gavarres	Marinas	Catalonia	Valencia	Adriatic	Mykonos	Corsica
Case-study setting										
Context of tourism development	nfl	nfuC	nful	nfl	bC	nbC	nbC	nbC	nuC	nuC
Scale of case-studies	L	SR	SR	L	T	R	R	R	SR	R
Present development										
Transformation of natural environment	1	1	2	1	3	3	3	3	2	1
Density of long-haul tourism flows	1	1	0	1	1	3	3	3	3	2
Density of recreation/day-trips	3	3	1	2	3	3	1	3	1	0
Seasonality of demand	3	3	1	2	1	2	2	3	3	3
Conflicts										
In tourism environment	2	1	2	3	0	3	3	3	2	2
Of sociocultural origin	0	0	0	3	0	0	0	0	0	3
In economic viability	1	1	2	0	1	2	3	3	1	0
Policy/planning structure										
Policy/planning responsibility	RL	NLP	NL	LP	LN	LRN	LRN	RLP	NPL	N
Integration in general planning mechanisms	3	3	1	1	3	1	2	1	0	0
Development restrictions	3	3	3	1–3	3	1–3	1	1	2	1–2
Sustainability										
Sought in study area	V	V C	V C	V S	C	C V	C	V C	V C	S V
Achieved to date	2	3 3	0 0	2 1	3	2 1	2	0 1	2 2	1 2
Future feasibility (by author)	3	3 3	2 2	2 2	3	3 2	2	1 2	3 3	1 2

Context: n, natural; f, farmland; u, some urban development; b, built-up; C, coastal; I, inland.
Scale: R, region; SR, subregion; L, local; T, site.
Levels: 3, high; 2, medium; 1, low; 0, none.
Agents: U, European Union; N, national; R, regional; L, local; P, private/public.
Sustainability: V, environmental; S, sociocultural; C, economic.

The case-studies have raised a range of issues concerning the rate, intensity and timing of development. No area has escaped some level of transformation of its natural environment but it is along Mediterranean shores that changes have been most intense, as a result principally of long-haul tourism, aggravated in some cases by demands generated from within the region itself. It is here that the influences of differential time-scales can best be seen, as reflected, for example, in the extent of development on the Catalan coast, largely exploited before sustainability issues were raised, and more recent initiatives on Mykonos, where there is evident respect for traditional urban and rural landscapes. The evolution over time should also be placed in a broader perspective. In the future, areas now carefully preserved may be engulfed in the urban hinterland of nearby cities under the pressure of urban expansion. This could occur in the case of Loch Lomond, for example, while, on the other hand, the cases of the proposed National Forest (Chapter 4) and the Gavarres forest area (Chapter 5) are a demonstration of the contrary situation whereby attempts are being made through legislation and planning to avoid or at least control urban expansion and sprawl, and to reserve certain areas as rural land.

Finally, a further dimension of tourism raised is the high seasonality of demand, with the obvious exceptions of urban-based marina sites and the proposed National Forest.

Conflicts Created

Throughout the text it is clear that tourism impacts take different forms, notably environmental, sociocultural and economic, which have been well identified in existing literature. By its very nature, the complexities of tourism inevitably result in a variety of tensions to which, since the late 1980s, the issue of sustainability has added a further potential dimension for conflict. However, the degree of conflict of the diverse impacts is, in fact, subject to different interpretations. An obvious distinction can be drawn between the attitudes of developers and those of conservationists, but at public authority level, among academics and within local populations similarly contrasting opinions can also be encountered. For example, for many Corsicans the mere presence of tourists is a source of antagonism (Chapter 11), whereas on the island of Mykonos tourist development has been harmoniously integrated into existing sociocultural and urban structures, since the island society has proved receptive to outsiders and the scale of tourist-related development has been strictly controlled. Even within the same region attitudes can vary, as is evident in Catalonia where, for example, the permissive and welcoming attitude towards large-scale tourist development in Lloret de Mar contrasts with the protectionist movement seeking to preserve the nearby Gavarres forest area.

The major preoccupations in the case-study areas are of an environmental nature, understood in terms of not only the natural but also the built environment. An initial presumption would obviously be that natural environmental issues were likely to be dominant but, in fact, it is precisely in built-up areas that high levels of conflict are being encountered. This would appear to imply that as tourism develops conflicts increase, but there are three exceptions in the case-studies to this generalization: the marinas, which are regeneration projects, the Gavarres forest in Catalonia, which is threatened by development, and the National Forest, which is an initiative aimed at promoting the use of an area for tourism while, at the same time, conserving it.

Economic viability is a preoccupation of those areas where mass tourism constitutes an important element in the local or regional economy, specifically those which have suffered some form of crisis in recent years – the northern Adriatic and Valencian coasts exemplify this. This crisis has been less ubiquitous in Catalonia, where intraregional tourism is also important. In many cases conflict is seen only to exist when an economic crisis arises or an impending environmental disaster looms.

Strategies Applied: Case-study Evidence

The case-studies reveal a basic distinction between strategies aimed at achieving sustainable development applied in intensively built-up areas which have been or are becoming tourist destinations and those which remain largely in their natural state. Obviously the type and extent of development prior to the awareness of sustainability issues is an important factor in the design of strategies. Although different strategies have been applied in different conflicting situations they reveal certain common characteristics.

In built-up destination areas, e.g. the UK marinas, Catalonia, Valencia and the Italian Adriatic, emphasis is placed on achieving sustainability, interpreted as long-term economic viability – or the durability of tourist activity – in an attempt to ensure that local inhabitants continue to at least maintain and preferably enhance their standards of living and disposable income. Such destinations have already undergone a cycle of development in some form, which may or may not be tourist-based. There is general recognition among those involved that these objectives can be achieved only within an increasingly attractive built environment. Under these conditions, urban regeneration involves soft landscaping and a renewed appreciation of traditional architectural styles and urban forms.

In areas which are in a more natural state, or where agricultural uses predominate, a major issue is the difficulty of striking an appropriate balance between use for tourism and recreation and the avoidance of the degradation of rural landscapes. Some strategies place emphasis on the maintenance of the status quo, as in the Gavarres forest. However, where degradation has taken place, the primary objective is the recovery of former landscapes, such as the sand-dunes of the Gower Peninsula (Chapter 3) and parts of the shores of Loch Lomond. The National Forest proposal is unique in that it constitutes an attempt to enhance an area, precisely to attract tourism and recreational activities.

Conflict Resolution

Before evaluating the degree of success in achieving sustainability through the strategies applied to overcome the conflicts identified, the meaning of success must be defined. This presents certain difficulties since success is subjective in three respects: in the first place, sustainability has been seen to have different interpretations; secondly, not all agents involved – developers, local citizens, pressure groups, public authorities, etc. – have similar expectations; and, finally, the tourism product, complex and multiform, is, moreover, subject to modification. Therefore success has to be evaluated from the point of view of these three variables, which are in themselves variable.

Reference has been made to the different forms of sustainability (Chapter 1) and to the different expectations of the agents involved in the case-studies (Chapters 2 to 11). Trends in demand indicate that the tourist consumer in general has become more discerning in evaluating standards of quality in both natural and built-up areas. As a result product sustainability depends on fulfilling these expectations in the different tourism environments. This is not the case where environmental sustainability is involved, as more objective parameters can be defined. Relevant examples include: soil erosion around Loch Lomond, sand-dune regression on the Gower Peninsula, forest depletion in the Gavarres, water pollution in the northern Adriatic Sea and coastline transformation on the islands of Mykonos and Corsica.

Within the range of measures introduced it is possible to identify a spectrum of shorter- and longer-term time horizons. There is evidence that individual actions of immediate application, which in themselves cannot be classified as long-term, tend to fulfil general long-term objectives of achieving sustainability when added together. The role of planning and policy-making at different levels is therefore important, as long-term objectives can be achieved through the integration of individual initiatives. The northern Adriatic Sea, for example, has registered a general improvement in water quality, particularly in relation to the problem of eutrophication in coastal waters. This is probably as a result of the implementation of diverse measures, including a reduction of the phosphorus content in detergents, controls on the use of nitrogenous fertilizers in farming, a reduction in discharges from industrial plants located on the coast, and the installation of improved sewage and water treatment plants in urban centres. In urban areas, minor regeneration projects aggregate to constitute larger-scale urban renewal, for example, as in the two UK marinas or many of the Spanish coastal resorts.

Administrative Structure and Feasibility

Each of the case-studies has pinpointed the issue of responsibility for dealing with the problems created by tourism development. Problem solution necessarily involves policy decisions, legislative measures and planning, as has been pointed out in Chapter 1. Conventional thinking recognizes a fourfold hierarchical structure of decision-making and policy implementation: supranational (European), national government, regional administrative bodies and local authorities. To these, however, must be added the increasingly important role of issue-orientated pressure groups, including, especially, environmental–ecological organizations and neighbourhood associations. The relative importance of each of these to individual case-studies is indicated in Table 12.1.

The case-studies show that there is no universal, generalized blueprint for the importance and power of each of these hierarchical components, as is illustrated by comparing Spain and the United Kingdom (Fig. 12.1). In the United Kingdom, all legislation and planning procedures emanate from the central government through the London Parliament while local democratically elected authorities – counties, districts – have very few innovative

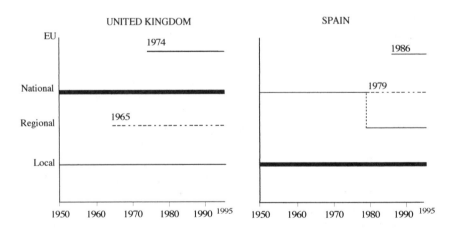

Fig. 12.1. Hierarchical structure of planning and policy-making in the United Kingdom and Spain. NB: width of line is related to degree of planning responsibility.

powers, which are, nevertheless, subject to central government overview. Regional government is absent, although a form of regional administration has existed since 1965 in Scotland and Wales – Scottish Office, Welsh Office – although not in England. In Spain, in contrast, planning and policy initiatives at national level have largely been lacking while municipal authorities have always had freedom to manoeuvre (Chapters 5 and 7). Following the introduction of regional government – autonomous communities – in Spain after 1978, some legislation and control at a regional level has been introduced to a varied extent according to each region. Throughout Europe, however, the influences of the European Union, directly through legislation and indirectly through the provision of grants for development and international cooperational initiatives, such as *The Blue Plan* for the Mediterranean (Batisse and Grenon, 1989) or trans-frontier agreements, are increasingly important.

The structure of tourism development evolves within the framework of existing planning and policy measures. There is strong evidence of a marked contrast between south European/Mediterranean and north European case-studies, both in the degree of integration of tourism in general planning systems and in the level of development restrictions associated with planning (Table 12.1). Further, the degree of integration and the level of restriction show a high degree of symmetry. Throughout Europe the role of public authorities to mediate between extreme positions is generally accepted. Nevertheless, where decisions made are perceived as being excessively permissive, alternative pressure groups can emerge to question and seek to modify those attitudes of authorities which are not necessarily regarded as publicly acceptable. There is evidence of this phenomenon in Palamós (Chapter 7), in the Gavarres forest – both in Catalonia – and in Corsica.

Questions which arise from this study include the extent to which the existing levels of tourism development depend only on the possibilities afforded within the framework of previously existing administrative structures, and the extent to which modifications to these are introduced as conflicts arise. Generally speaking, it would appear that, where no well-developed planning structure existed prior to tourism development, it is only in a situation of extreme economic or environmental crisis that such a structure is created or modified (Chapter 8). On the other hand, crises can arise even within long-established and strictly defined structures (Chapter 3).

Answers to these questions may possibly be found by examining the sustainability objectives of the different cases and the level of success achieved, which may permit a forecast to be made (Table 12.1). Given that the principal conflicts identified were of an environmental nature, it is not surprising that sustainability objectives were in this sphere. Where economic viability also emerged as a priority, a high-quality tourism environment – natural and/or built – was seen as essential to achieve these objectives. Competitiveness in tourism is obviously related not only to price but also to

environmental quality. The level of achievement of the goals is particularly
high in the Gower Peninsula, where strict planning controls have been exercised for many decades, and in the urban marina developments where planning and development has been integrated and holistic from project incep-
tion. In like vein, the tourism environment in Mykonos, which is perceived
as being of high quality by visitors, is a major factor in maintaining demand.
In the light of such examples, future feasibility would appear to depend on
integrated planning, as proposed by Coccossis in Chapter 1, and as argued
by Dickinson for Loch Lomond, Ballinger for the Gower Peninsula, Nicholls
for the National Forest, Priestley for Catalonia, Vera and Rippin for Valencia
and Zanetto and Soriani for the northern Adriatic Sea.

General Perspectives

There is a clear need to make a distinction between the sustainability of a
particular tourist product and the sustainability of the environment. In some
cases interest in the two can coincide, especially where the natural environment is the main element of attraction. However, where the natural environment has already been greatly transformed and/or degraded, regeneration
is the principal necessity whether it is in a rural or urban location.

The key to sustainability in all cases is that the product offered is maintained in such a way that it satisfies consumer expectancies and therefore
demand levels are maintained on a long-term basis. These are identified and
expressed in the model of tourism environment expectancies for sustainable
tourism (Fig. 12.2), which provides a framework for determining thresholds
for sustainable tourism development in the future. However, this should not
be considered a static model for, as consumer standards change, the model
will need to be reviewed and revised.

The model incorporates four elements: the degree of transformation of
the natural landscape (vertical axis); the time horizon (horizontal axis),
which offers a projection into the future from the time at which sustainability objectives are established; the three key elements of tourism
which are subject to impacts – natural areas, sociocultural traits, built-up
areas; and the range of expectations – from acceptable to ideal – which
permits sustainable tourism development. The model, therefore, identifies the
basic distinction already made between tourism products based on natural
landscapes and built-up environments but also incorporates, as a distinctive
element, sociocultural aspects. Clearly, in the case of tourist products based
on natural landscapes, further degradation or simply transformation is inad-
missible and, ideally, policies aimed at restoring the natural landscape should
be implemented. In a built-up environment, the restoration of certain
elements can be sufficient to satisfy expectations and to provide a basis for
new dimensions of sustainable tourism development. Modifications of socio-

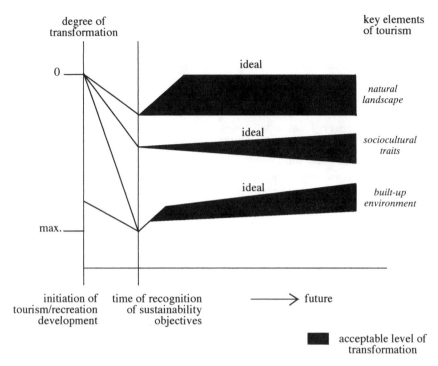

Fig. 12.2. Model of tourism environment expectancies for sustainable tourism development.

cultural traits appear inevitable as external influences of a general nature, e.g. mass media, inexorably join tourism in generating cultural changes which have global characteristics, e.g. music, dance, dress, food.

In the context of a spatially expanding European political framework, a large part of tourism and recreation demands in the forseeable future will be directed towards two market sectors: intraregional leisure and recreation activities to satisfy the needs of local populations, and long-haul tourism concentrated along the pleasure periphery of Europe. In both cases, the success of the tourism product will depend on its capacity to satisfy the expectancies of an essentially urban-based clientele which is geared to expect rising standards of environmental quality.

Finally, while a distinction between the natural and built environments can be made, sustaining the tourism product in both requires constant regeneration. And, while the products are very different, both types must be sustained in such a way that they satisfy expectancies in order to maintain demand. Throughout the range of case-studies, there is evidence that the present and future development of tourism requires a sustainable dimension. Achieving such goals needs a diversified rather than a unified strategy, in which long-term objectives must be a priority. The common message that has emerged is the need for coordination among the different agents involved in order to achieve a degree of sustainability. This does not necessarily involve rigid regulation; rather, there must be sufficient flexibility in planning procedures to enable them to adapt to changing trends and circumstances. Moreover, local involvement is fundamental to planning and management, and in the acceptance of proposals that are of concern to the general public. Such linkages with society are necessary, at the sources of origin of demand but especially at tourist destinations. Sustainability will not be easy to achieve in tourism, but it is hoped that experiences such as those analysed in this book will help to identify some strategies for the future.

References

Batisse, M. and Grenon, M. (eds) (1989) *Futures for the Mediterranean Basin: the Blue Plan*. Oxford University Press, Oxford, 279 pp.

Author Index

Place Index

Subject Index